Organic Stereochemistry

Henri Kagan

Professeur à l'Université de Paris-Sud

**Translated by M. C. Whiting
and U. H. Whiting**

A Halsted Press Book

**John Wiley & Sons
New York**

© Edward Arnold (Publishers) Ltd 1979

Authorized translation from the French
La stéréochimie organique
published by Presses Universitaires de France, 1975

This translated edition first published in 1979 by
Edward Arnold (Publishers) Ltd
41 Bedford Square, London WC1B 3DQ

Published in the U.S.A.
by Halsted Press, a Division
of John Wiley & Sons, Inc.
New York

Library of Congress Cataloging in Publication Data
Kagan, Henri.
  Organic stereochemistry.

  Translation of La stéréochimie organique.
  "A Halsted Press book."
  Includes index.
  1. Stereochemistry.   2. Chemistry, Physical organic.
I. Title.
QD481.K2513   547'.1'223   79–11500
ISBN 0–470–26725–9

Printed in Great Britain

Preface

Stereochemistry constitutes an area of chemistry itself, not a distinct science; but it is an area conveniently treated as a unit. Not long ago one would not have needed to prefix the title with the word 'organic', since stereochemistry was born within organic chemistry, and permeated by the ideas of organic chemistry. With the development of effective methods of studying the solid state, however, inorganic chemistry has developed a distinct and comparably large stereochemistry of its own, and is now better treated separately. The present edition of a book first published in France appears well suited to the education systems of the United Kingdom and the United States, making few assumptions about previous knowledge, but reaching a level of understanding that should suffice for any graduate student.

Stereochemistry derives from the work of Pasteur, perhaps more completely than chemistry itself from the work of Lavoisier. It was in French that van't Hoff wrote 'La chime dans l'espace', a book that in 1875 brought his and Le Bel's appreciation of the tetrahedral carbon atom and its consequences to the attention of a rather sceptical world; a book that would have served adequately as a textbook of stereochemistry until, say, 1950. Contact with the continuing French tradition in stereochemistry could be valuable to an English-speaking chemist, and the translators do not apologise because the style and language of the original can be sensed in this often rather literal translation. They have, however, taken the chance to correct a few unimportant slips and misprints, and to add a few footnotes.

Some of the matters dealt with in this book, particularly conformational analysis and stereoisomerism, were dealt with in the *cours de D.E.A.* at the Université de Paris-Sud. We thank R. Bucourt, J. Weill-Reynall and C. Mazieres for their valuable advice in the editing of the French text.

1978

H B K
M C W
U H W

Contents

Introduction

In the definition of stereochemistry it is helpful to remember that the prefix 'stereo' originates from the Greek root *stereos*, solid or volume. Stereochemistry is, therefore, 'chemistry in space'. This definition is that which was given by Le Bel and van't Hoff in 1874 when these authors advanced their famous hypothesis of the tetrahedral carbon atom. Such a definition would be completely erroneous if it were held to imply the existence of two chemistries, of which one was spatial. Chemical reactions, which imply an approach of reactants and the formation of a transition state, all take place in space in three dimensions. The molecules of which we are considering the structure are not completely defined by a planar formula, and it is necessary to specify the relative position of the atoms, one to another. Because of the progress of chemistry, stereochemistry and chemistry have become inseparable. Chemistry 'on paper', i.e. in two dimensions, making use only of planar structures, is completely inadequate when a chemical phenomenon, for example a reaction mechanism, is examined in detail. Molecular models and their symbolic planar representations are of great value to the chemist for predicting isomers and understanding reactions.

Stereochemistry thus places the emphasis on the steric aspect of chemistry, and valuable information is obtained by examining a reaction, however simple, from the point of view of stereochemistry.

Stereochemistry has played an important role in the development of chemistry. Crystallography enabled the chemists of the last century to become familiar with crystalline forms, which are related to the organization of the crystalline lattice and indirectly to the structure of the molecule. In 1848, Pasteur, due to his understanding of crystallography and his powers of observation, succeeded in separating two types of crystal of the sodium ammonium tartrate formed from racemic tartaric acid; one the mirror image of the other. This first resolution of a racemic compound made the world aware of a particularly important form of isomerism, optical isomerism. Thirty years later, Le Bel and van't Hoff made the optical isomerism of compounds

$$R_1 \diagdown \quad \diagup R_2$$
$$C$$
$$R_3 \diagup \quad \diagdown R_4$$

their main argument in eliminating the possiblity that the four valencies of carbon were coplanar. In the particular case of carbon, structural theory was thus placed on a solid foundation. The arrangement of valencies around an atom of nitrogen was similarly studied by stereochemical methods. It was found that quaternary ammonium cations

$$R_1 \diagdown \quad \diagup R_2$$
$$\overset{+}{N}$$
$$R_3 \diagup \quad \diagdown R_4$$

and amine oxides

$$R_1 \diagdown$$
$$R_2 \!-\! \overset{+}{N} \!-\! \overset{-}{O}$$
$$R_3 \diagup$$

could be obtained optically active. In these compounds therefore, nitrogen does not have a planar structure. This result was particularly interesting for studies on the semipolar bond $\rightarrow\!\overset{+}{N}\!-\!\overset{-}{O}$. Compounds of sulphur, phosphorus and arsenic were also examined by these techniques.

The chemistry of metallic complexes made considerable progress at the beginning of the century as a result of the work of Werner, for which the Nobel Prize was awarded. Werner put forward the revolutionary hypothesis of octahedral cobalt or nickel in complexes where these metals were hexacoordinate. The metallic atom is at the centre of an octahedron and the ligands are placed at the corners. This representation of chemical bonds around a metal has been particularly fruitful, and has allowed the prediction of all the isomers possible as a function of the nature of the ligand L.

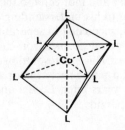

There are many additional examples which demonstrate the essential role of stereochemistry in the evolution of the concepts of chemical bonding and reaction mechanisms.

Stereochemistry permeates all chemistry, whether it is described as inorganic, organic or physical. The frontiers between disciplines are difficult to define, and it is in any case undesirable to create divisions which can soon become sterile. There has recently been a rediscovery of inorganic chemistry by organic chemists, who have learned to carry out some remarkable organic syntheses by using catalysts based on organometallic complexes. Physical chemistry has yielded methods for studying molecules and their stereochemistry, and for throwing light on transient reaction intermediates. The use of physical methods in the last decade has revolutionized organic chemistry and profoundly changed its character.

As the aim of this book is to emphasize organic stereochemistry, its scope must first be defined. Inorganic chemistry will not be considered, although this has an important stereochemical aspect; the chemistry of boron, phosphorus, sulphur and transition metal complexes, for example. The book will be limited to conventional small organic molecules composed of carbon, hydrogen and various heteroatoms (oxygen, nitrogen, sulphur and phosphorus are among the most frequent), although organic macromolecules and the stereochemical problems posed by biochemistry will also be examined superficially. It must, however, be noted that, in these latter branches of organic chemistry, structural and stereochemical aspects play an increasingly important role.

The apparent division of chemistry has been emphasized. This division has become more and more evident, because of the importance of experimental methods which are related to the nature of the compounds studied. Naturally the methods one uses vary according to whether one is trying to isolate a gas, a water-soluble compound (e.g. a salt or an amino acid) or a compound soluble in an organic solvent, or whether one wishes to study the chemistry of the solid state. But this great diversity of chemistry fortunately does not imply an extreme specialization for the chemist himself, if he can raise himself above the descriptive and technical aspects, which are important but not essential. For example, the conformational analysis of hydrogen peroxide, $HO—OH$, or of ethane, $H_3C—CH_3$, and of their derivatives, proceeds by the same principles and makes use of the same concepts.

In all branches of chemistry, it is evident that there are two aspects of stereochemistry; static and dynamic.

Static stereochemistry deals with the structure of molecules, that is to say the relative positions of atoms in space. Frequently a molecule is not rigid, and its shape can vary according to the temperature or the nature of the substituents. These variations are generally due to rotations about single bonds and are studied with the aid of conformational analysis

Stereochemistry allows one to predict the number of stereoisomers corresponding to a planar formula, i.e. to a given sequence of atoms. Methods have been perfected for studying the stereochemistry of molecules. In the chemistry of natural products these methods are essential for the complete definition of the structure of an isolated compound. The interest of such a determination is not purely academic; the repercussions are considerable if the substance possesses therapeutic activity and knowledge of its structure makes a partial or total synthesis possible on an industrial scale. Cortisone is an example; there were only a few milligrams of cortisone in the world in 1945, its anti-inflammatory action was discovered in 1947, and its partial synthesis from the bile acids was carried out in 1950 on a kilogram scale.

Dynamic stereochemistry deals with the reactivity of molecules. Understanding the steric course of a reaction implies understanding the stereochemistry of the products formed and, if possible, the course of the reaction and the geometry of the transition state. Stereospecificity should be a characteristic of a modern organic synthesis. It always constitutes a difficult problem, which belongs to the field of dynamic stereochemistry. As an example of the importance of steric control in synthesis, it can be calculated that a compound of which the planar formula contains three asymmetric carbon atoms, for example penicillin, can exist in the form of $2^3 = 8$ distinct stereoisomers, but only one isomer possesses the desired antibiotic activity. The study of dynamic stereochemistry is a necessary means for obtaining precise information on reaction mechanism (ionic, radical, photochemical and even enzymatic reactions). A well-known example is the Walden inversion (see Chapter 5). As early as 1903 Walden had shown while working with optically active molecules that in certain cases of nucleophilic substitution, now called S_N2, an inversion of configuration occurred which implied a reversal of the initial tetrahedron:

In this book organic stereochemistry will be developed progressively. In Chapter 1 the electronic properties of atoms which constitute organic molecules will be revised. These comprise the basic parameters (intervalency angles, interatomic distances, orbitals, etc.) which to a large extent constrain the geometry of the molecule.

Chapter 2 presents molecular models and methods of representing molecules.

Conformational analysis, described in Chapter 3, allows molecular structures to be analysed when there are possibilities of rotation around certain bonds. The shape of a molecule is then susceptible of considerable variation; some conformations are much more stable than others, and chemical reactivity is related in most cases to molecular conformation.

In Chapter 4, different types of stereoisomerism are reviewed. Finally, dynamic stereochemistry is treated in Chapter 5.

We hope that at the end of this work we will have succeeded in giving the reader some insight into one of the recent trends in modern organic chemistry.

1

Molecular geometry and chemical bonding

It is usual to define organic chemistry as the chemistry of carbon compounds. Organic compounds always contain carbon and usually hydrogen; very often heteroatoms such as oxygen, nitrogen, sulphur, phosphorus, the halogens or alkali metals are also present.

The great wealth of organic chemistry has several causes. The stability of covalent C—C and C—H links is an essential factor, allowing the synthesis of complex chains of carbon atoms. The quadrivalence of carbon permits structural isomerism by multiplying the possible arrangements of atoms. The non-coplanarity of the valencies issuing from a tetragonal carbon atom, and the planarity of an ethylenic double bond, give rise to stereoisomers. Stereochemistry begins with the characteristic properties of the linkages formed between the atoms which constitute a molecule. Before considering the geometry of simple molecules it is therefore necessary to discuss the nature of chemical bonds.

Revision of the theory of chemical bonding

The bond formed between atoms makes use of their outer electrons. It is no longer conceivable to consider the problem of the chemical bond without a quantitative treatment of the energies and spatial distributions (or their equivalent) of the molecular orbitals occupied by the bonding electrons. It will be dealt with here by a simplified approach, which will suffice for the many structural problems of organic chemistry.

Electrons are arranged around each atom in shells (K, L, M, etc.), each shell being divided into subshells (s, p, d, f), themselves subdivided into orbitals containing at the most two electrons of opposed spin.

To determine the number of outer electrons of an atom it is necessary to know the total number of electrons of that atom (equal to its atomic number Z) and the rules for filling electronic shells. These rules are based on the fact that an electron, characterized by four quantum numbers, must be different from all other electrons (the Pauli principle). The principal quantum number n defines the shell: $n=1$, shell K; $n=2$, shell L, etc. The secondary quantum number l differentiates the subshells and can take the values between 0 and $(n-1)$: $l=0$, subshell s; $l=2$, subshell p, etc. The magnetic quantum number m (possible values from $-l$ to 0 to $+l$) introduces the division into orbitals in which the electron can have spin states $+\frac{1}{2}$ or $-\frac{1}{2}$ (indicated by ↑ or ↓ to symbolize electrons with rotations in opposite directions.

Consider, as an example, the electronic configuration of a carbon atom $(Z=6)$, assumed to be 'free'. The most stable orbitals are filled first, that is to say going outward progressively from the nucleus. In Figure 1.1 the distribution of the six electrons is represented in two equivalent ways. Note that the p electrons are situated in two orbitals rather than in the same orbital, according to Hund's rule, which predicts greater stability for this distribution.

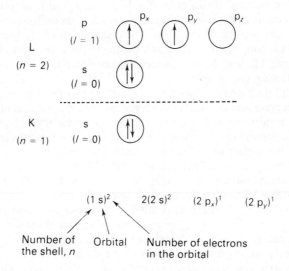

Figure 1.1   Electronic configuration of the carbon atom in its ground state

In discussing the chemical bond only the outer electrons of the atom need be considered. As a result of the behaviour of these electrons, bonds can show all characteristics intermediate between the two extreme cases which are commonly called electrovalence and covalence.

The electrovalent bond is common in inorganic chemistry. Conceptually it is very simple: electrostatic attraction binding two ions of opposite charge. These ions are formed by the exchange of one or more electrons between two atoms. The atom which loses an electron becomes a cation, the atom which accepts an electron becomes an anion.

In the covalent bond each atom is at the same time a donor and an acceptor of an electron. The bond is constituted by the sharing of two electrons, which become the doublet of the bond.

The diagrams of G. N. Lewis give a simplified representation of ions and molecules. In these diagrams only the peripheral electrons are considered, grouped, when possible, as doublets.

Electrovalence ClNa $:\!\ddot{Cl}.$ $^\times$Na $\longrightarrow$ $[:\!\ddot{Cl}\!:]^{\ominus}$ [Na] $^{\oplus}$

Covalence Cl_2 $:\!\ddot{Cl}$ $\ddot{Cl}\!:$ $\longrightarrow$ $:\!\ddot{Cl}\!:\!\ddot{Cl}\!:$ ou Cl—Cl

Very often the transfer of electrons leads to atoms surrounded by an octet of electrons, electronic structure s^2p^6. An octet of electrons is a factor for stability, which in particular characterizes the rare gases. Cl^- and Na^+ are surrounded by an octet (for Na^+ there is also an inner shell not represented). In Cl_2 one can equally consider that each chlorine atom contains eight peripheral electrons. The distribution of electrons in the L shell of carbon, Figure 1.1, allows us to predict a bivalent character for this atom. The well-known quadrivalence of carbon will be explained in the following paragraphs.

Quantum mechanics shows that the particle theory of the electron is incomplete; one cannot specify with certainty both the position of an electron and its momentum. It is wrong to think of covalency as represented by two electrons, placed between two atoms and shown as dots in Lewis's diagrams. It is nearer to reality to represent the electron as an electronic cloud of total charge $-e$. This cloud has a density which varies according to the position considered and this electronic density, or the density of the probability of finding the electron, is equal to $|\Psi_{(x,y,z)}|^2$.

$\Psi_{(x,y,z)}$ is the wave function associated with the electron, also called the orbital. A knowledge of $\Psi_{(x,y,x)}$ gives us an indication of the shape of the electronic charge in space. Calculation shows that an s orbital has spherical symmetry, while the three p orbitals have the symmetry of revolution around an axis, the axes being perpendicular to each other. The size of the electron cloud increases as shells become successively more distant from the nucleus, but the shape remains unchanged. Starting from the functions Ψ it is possible to plot the electronic density by calculating $|\Psi|^2 dv$. The greatest part of the electronic cloud, particularly the regions of high density, is localized in a volume having essentially the same symmetry as that of Ψ. Figure 1.2 shows a simplified but convenient representation of the electronic clouds s and p (called, incorrectly, orbitals). The sign of Ψ allows one to introduce the consideration of symmetry, which is one of the main features of present-day structural chemistry.

The covalent bond is discussed in quantum mechanics in terms of the overlap of atomic orbitals. Some stereochemical consequences are immediately predictable.

Figure 1.3 shows different cases of overlap of atomic orbitals. In every case a linear combination of orbitals Ψ_A and Ψ_B gives two possible solu-

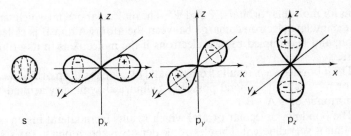

Figure 1.2   Schematic representation of orbitals (calculated from $|\psi_{(x,y,z)}|^2$)

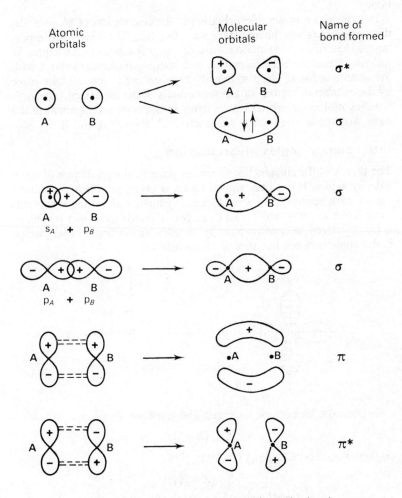

Figure 1.3   Overlap of atomic orbitals: formation of σ and π bonds

tions for molecular orbitals: Ψ and Ψ^*. The molecular orbital which tends to accumulate electronic charge between the atoms A and B is called a bond, and is occupied by two electrons if the molecule is in its ground state.

The σ bond always results from an axial overlap for the p orbital, $(s_A + p_B$ or $p_A + p_B)$. The σ bond possesses cylindrical symmetry around the interatomic axis A—B.

The bonding molecular orbital which results from lateral fusion of p orbitals is very different. The electronic density is zero along the axis A—B, the electronic cloud being concentrated on either side of this axis (π bond).

In diagrams representing orbitals (the representation of Ψ or $\Psi^2 dv$) it is essential to show the sign of the wave function. These signs are important in determining the steric course of many reactions (*see* Chapter 5). Calculations require the formation of a bonding molecular orbital, with the accumulation of electron density between the atoms, to take place by the overlap of regions of atomic orbitals of the same sign. The antibonding molecular orbital results from an opposite arrangement of the signs, and tends to diminish the function Ψ between A and B.

Intervalency angles, hybridization

The theory of the chemical bond makes possible the prediction of intervalency angles. It has been shown that a σ bond, in which a p atomic orbital participates, possesses rotational symmetry about the axis of this orbital. An atom A which can form two σ bonds from two p orbitals is characterized, in consequence, by an intervalency angle of $90°$, which is the angle between the axes of two p orbitals.

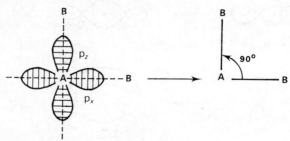

Let us take, as examples, oxygen and sulphur:

$$O \quad (Z=8)$$

configuration $(1s)^2(2s)^2(2p_x)^2(2p_y)^1(2p_z)^1$;

$$S \quad (Z=16)$$

configuration $(1s)^2(2s)^2(2p)^6(3s)^2(3p_x)^2(3p_y)^1(3p_z)^1$.

The valency electrons are in the 2p and 3p orbitals respectively. This leads to the bivalency of these elements and to the prediction that the two bonds will be orthogonal to each other. In the molecules H_2O and H_2S the angles have been determined experimentally:

The oxygen atom thus shows unexpected behaviour, which can be interpreted in different ways. A first suggestion was a repulsion between the hydrogen atoms, leading to an angular deformation away from the theoretical value of 90°. A more reasonable explanation is based on a changing of the character of the atomic orbitals of oxygen. Instead of having a pure p character they are affected by the 2s orbital. As an extreme view, one can imagine a complete mixture, a 'hybridization' between s and p orbitals, leading to the formation of new orbitals, all equivalent. The concept of hybridization is only an approximation, but one still extremely useful for organic chemists although largely replaced by a broader treatment (molecular orbitals making use of all the atomic orbitals of the atoms involved in the linkages). If hybridization involves an s orbital and 3p orbitals, the four new molecular orbitals which result are called sp^3. Calculation shows that each sp^3 orbital has a symmetry of rotation (*see* Figure 1.4), the four axes which correspond forming between them angles of 109° 24'. Two single bonds formed from two sp^3 orbitals retain, in principle, the angle of 109°. The angle of 105° found for water could thus be an indication of marked sp^3 hybridization for oxygen.

The hybridization of one s and two p orbitals results in sp^2 hybrid orbitals. The axes of these orbitals are coplanar, each making an angle of 120° with the other two. The hybridization of one s orbital and one p orbital gives two sp orbitals, of which the axes are collinear. The mathematical treatment which allows the definition of hybrid orbitals consists of the linear combination of atomic orbitals. Each hybrid sp^{λ^2} atomic orbital can be represented by the linear combination $s + \lambda$ p, where λ is the parameter of hybridization. If the orbital has a p character of 50% (sp orbital), $\lambda^2 = 1$, and the angle between the axes of the two sp orbitals is 180°. An intervalency angle of 90° corresponds to pure p character ($\lambda^2 = \infty$). An increase in the s character of an atomic orbital results in an increase in the intervalency angle. So far only particular values of λ have been considered, for example $\lambda^2 = 3$, giving sp^3 orbitals. One can generalize the idea of hybridization to any value of λ, which will explain the examples where intervalency angles are different from the particular values mentioned above.

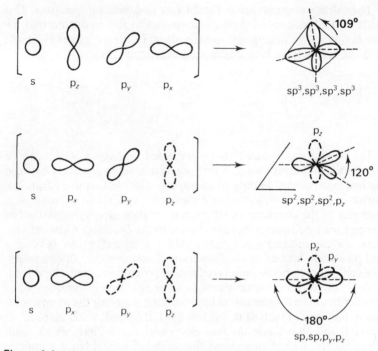

Figure 1.4   Hybridization of s and p atomic orbitals

It has been assumed implicitly that all intervalency angles coming from the same atom are equal, as a result of the equivalence of the hybrid orbitals. This amounts to an approximation which is true only for molecules in which all the valencies of an atom are effectively equivalent (BF_3, CH_4, etc.). Less symmetrical molecules will be discussed later.

Bonding in carbon compounds

The electronic configuration of the carbon atom was given in Figure 1.1, allowing a bivalent character to be predicted for it. In the large majority of cases, however, the experimental data prove that carbon is quadrivalent, the four valencies being more or less equivalent in saturated derivatives. In 1874, Le Bel and van't Hoff proposed that the carbon atom was at the centre of a regular tetrahedron, the bonds being directed towards the corners. The intervalency angles thus defined are 109° 28', which is comparable to those measured in saturated hydrocarbons.

Present theories of the chemical bond are particularly useful in explaining the various types of valency formed by a carbon atom. Apart from some exceptional cases there is partial or total hybridization of the orbitals 2p and 2s.

The linear combination of wave functions

$$\Psi_{2s}\Psi_{2p_x}\Psi_{2p_y}\Psi_{2p_z}$$

is favoured energetically and gives four sp^3 orbitals. The trigonal carbon atom present in olefins is sp^2 hybridized, leaving one of the p orbitals originally present. Finally, acetylenic carbon is a carbon atom with sp hybridization which allows two p orbitals to remain. Therefore, the electronic configuration of sp^3, sp^2 and sp carbon atoms is as follows:

Tetragonal C	$(sp^3)^4$
Trigonal C	$(sp^2)^3(2p_z)^1$
Digonal C	$(sp)^2(2p_y)^1(2p_z)^1$,

Figure 1.4 shows the various atomic hybrid orbitals and the orbitals which remain unchanged. To understand how the interatomic linkages are established so as to constitute molecules, one must begin from the principle that a linkage (bonding molecular orbital) is formed from the maximum overlap of two atomic orbitals.

Single bond, C—C   The axial overlap of two sp^3 orbitals is very favourable for forming a σ bond:

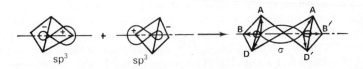

The σ bond has a symmetry of revolution such that rotation around the interatomic axes changes the relative positions of the substituents A, B, D without affecting the electronic distribution of the σ pair. It has been known for thirty years that rotation around a single C—C bond is possible, but is hindered in certain positions; the barriers to rotation will be discussed in Chapter 3.

Double bond, C=C   This is formed between two atoms having sp^2 hybridization. The best overlap between two sp^2 orbitals is of the axial type. On each carbon atom a p orbital remains available, and a new linkage can be formed by lateral overlap between the p orbitals. This overlap is possible only if the two carbon atoms which are bound by the σ bond are so orientated as to align the axes of their p orbitals in parallel, which has the effect of making the axes of the six sp^2 orbitals coplanar.

A C=C double bond is thus formed from one σ bond and one π bond. The π bond, which is composed of a double electronic cloud, completely prevents rotation round the σ bond and makes the double bond planar. Cis–trans ethylenic isomerism, known as geometric isomerism, originates from this circumstance. Consider the two isomeric but-2-enes:

$$CH_3 \diagdown \quad \diagup H$$
$$C = C$$
$$H \diagup \quad \diagdown CH_3$$

trans

$$CH_3 \diagdown \quad \diagup CH_3$$
$$C = C$$
$$H \diagup \quad \diagdown H$$

cis

Two nonsuperimposable molecular structures are possible.

Geometric isomerism is the simplest example of stereoisomerism. It is interesting to note that the planar structure of an olefin disappears when it is in the excited state, for example from absorption of ultraviolet radiation of suitable frequency. The extra energy gained by the molecule raises one of the electrons of the π pair to the empty molecular orbital π^* of greater energy:

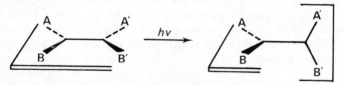

The terminal bonds tend to place themselves in two perpendicular planes.

Triple bond, C≡C The maximum overlap of atomic orbitals for two sp carbon atoms is obtained when the sp axes coincide, the axes of the p orbitals then being parallel to each other:

Acetylenic compounds $R-C\equiv C-R'$ are linear, the σ bond being surrounded by a π cloud of four electrons of cylindrical symmetry.

Allenes, $C=C=C$ This example shows again the value of considering the state of hybridization of the atoms in order to obtain information on the spatial structure of a molecule. Consider allene:

and ethylene:

It is useful to know whether the allene molecule is planar like the ethylene molecule.

The C_1-C_2 linkage of allene is formed with a carbon atom at 2 which is bi-coordinate and should have sp hybridization. For maximum overlap of atomic orbitals it is necessary that the axes of the p_z orbital of the two atoms be parallel. The C_1-C_2 linkage has all the characteristics of a double bond. Carbon atom 2 has two available orbitals. The sp orbital collinear with the C_1-C_2 axis will overlap axially with the sp^2 orbital of carbon atom 3, and the p_y orbital perpendicular to the plane of the double bond C_1-C_2 will combine with the p orbital of carbon atom 3 by lateral overlap. It is, therefore, necessary to place the orbital of carbon atom 3 perpendicularly to the plane of the diagram, which will orient the two bonding sp^2 orbitals in the plane of the diagram. It is clear that the σ bonds which will be formed at each end of the molecule will be in two planes, one perpendicular to the other. Experiment confirms that the allene molecule is not planar.

Structure of free radicals

A radical is a very unstable species $R_1R_2R_3C\cdot$, in which the carbon atom is tervalent and possesses a single electron. Some radicals have been isolated, but in the majority of cases they react as soon as they are formed. Free radicals have been studied by electron spin resonance (e.s.r.). Simple radicals like methyl, $CH_3\cdot$, are planar and the intervalency angle is close to 120°. The carbon atom is in a state of sp^2 hybridization, the lone electron occupying a free p orbital:

$$C : (1s)^2(sp)^2(2p_z)^1$$

Free radicals, based on carbon, in which halogen atoms are bonded to the carbon atom, are not planar, although they are considerably flattened by comparison with the tetrahedral structure.

Carbanions and carbonium ions

Ionic reactions frequently involve charged intermediates which are formed by heterolytic rupture of a bond:

$$CH_3 \overset{\frown}{} A \longrightarrow CH_3^- : \text{ (carbanion)} \quad + A^+$$
$$CH_3 \overset{\frown}{} A \longrightarrow CH_3^+ \quad \text{(carbonium ion)} + A^-$$

A carbanion is obtained from a radical by gaining an electron, whereas a carbonium ion results from the loss of an electron from a radical. A carbanion has a tetrahedral structure, the carbon atom being of sp^3 hybridization, the lone pair occupying an sp^3 orbital and playing the role of a substituent. An important difference between saturated hydrocarbons and carbanions should be noted. The carbon atom carrying the lone pair undergoes rapid inversion in carbanions, at the half-way point of the reaction being in the sp^2 hybridization state. Carbonium ions are planar, the positive carbon atom being in the sp^2 hybridization state and the p_z orbital being empty.

Carbenes R—C—R'

Carbenes are very reactive compounds which were already being studied by Doering in 1950. They are neutral, the central carbon atom being bivalent. The structure of the simplest carbene, methylene, CH_2, has been the object of many spectroscopic studies and calculations by theoretical chemists. Carbenes can exist in two electronic configurations; the triplet state is characterized by parallel spins for the two electrons not involved in bonds, whereas in the singlet state these two electrons possess anti-parallel spins. It seems to be adequately proved that triplet methylene is much less bent than singlet methylene, in which the intervalency angle is near to $103°$.

An approximate description of the two types of carbene is that of an sp hybridized carbon atom for the triplet state and an sp^2 carbon atom for the singlet state. Two electrons of triplet methylene occupy two p orbitals orthogonal to each other, whereas singlet methylene does not possess unpaired electrons; in it an sp^2 orbital is filled by a doublet, the p_z orbital remaining unfilled. These descriptions rationalize the different chemical properties of the two types of carbene which can be made in the laboratory. The triplet carbene has the character of a biradical, the singlet carbene has simultaneously carbanion properties by virtue of its electron pair, and carbonium ion properties due to its empty orbital.

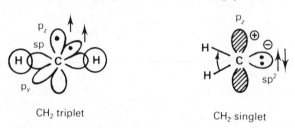

CH₂ triplet   CH₂ singlet

The chemical bonds formed by oxygen

Oxygen atoms are present in many functional groups. The oxygen atom is bound to carbon by a single bond in alcohols, $\geqslant$C—OH, ethers, R—O—R', and epoxides, $>$C$\overset{\displaystyle O}{\underset{\textstyle}{\diagup \diagdown}}C<$. Double carbon–oxygen bonds $>$C$=$O are also frequently found, notably in acids and their derivatives and in aldehydes and ketones.

The electronic structure of the oxygen atom in the ground state is $(1s)^2(2s)^2(2p_x)^2(2p_y)^1(2p_z)^1$.

It has been seen that even in a molecule as simple as H_2O the intervalency angle is not $90°$. In Figure 1.5 the intervalency angles of various oxygenated saturated molecules are shown. They are always considerably larger than 90°, which is evidence of an sp^3 state of hybridization for the oxygen atom.

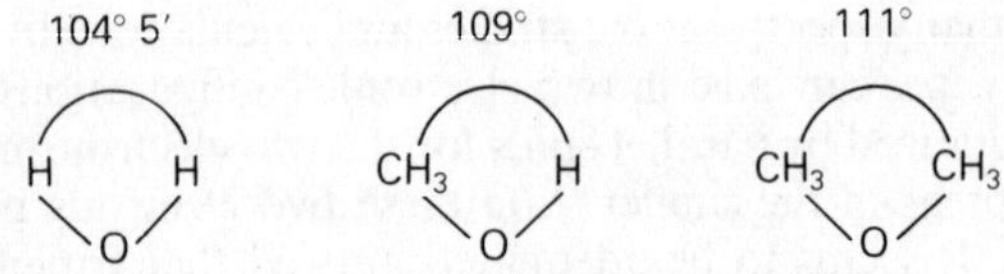

Figure 1.5 Intervalency angles of oxygen compounds

The chemical bonds formed by sulphur

Sulphur is present in many organic compounds.

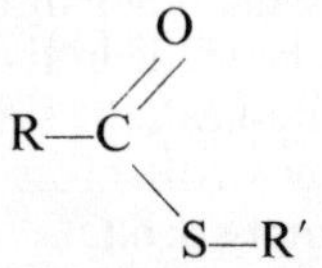

Thioester

R—S—R′

Thioether

RSH

Thiol

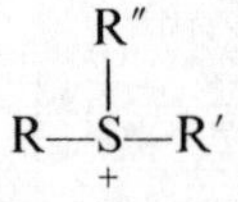

Sulphonium cation

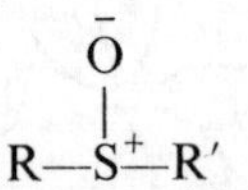

Sulphoxide

The electronic state of the sulphur atom ($Z=16$) is as follows:

$$(1s)^2(2s)^2(2p)^6(3s)^2(3p_x)^2(3p_y)^1(3p_z)^1.$$

In compounds R—S—R′ the intervalency angle is close to 90° (for example, 92° in SH_2). The sulphur atom thus uses its p orbitals. In tri-substituted derivatives of sulphur the sulphur atom is not planar. The case of sulphoxides is interesting. It may be suggested that the sulphur atom gives two electrons to oxygen, forming a semipolar linkage represented by the formula:

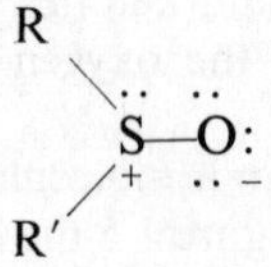

On the other hand, the formation of a double bond between sulphur and oxygen leads to the structure:

$$\begin{array}{c} R \\ \diagdown \overset{..}{}\overset{..}{} \\ S{=}\ddot{O}: \\ \diagup \\ R' \end{array}$$

The latter structure contradicts the octet rule for the sulphur atom, and is possible only by the use of the vacant 3d orbitals of sulphur, which accept the lone pair of electrons of the oxygen atom. The overlap between the p orbital of oxygen and the d orbital of sulphur, which leads to a π bond, is denoted p_π—d_π. It does not require planarity for the sulphur atom, and sulphoxides are not planar.

The chemical bonds in nitrogen compounds

Nitrogen is extremely important in biochemistry. Purine bases are present in the chains of nucleic acids, alkaloids are natural compounds widely distributed among plants, and polypeptides form the skeleton of enzymes and proteins.

The nitrogen atom is tervalent in amines, RNH_2, imines,

$$\begin{array}{c} R \\ \diagdown \\ C{=}N{-}R'', \\ \diagup \\ R' \end{array}$$ and nitriles, $RC{\equiv}N$, but it is quadrivalent in oxides,

$$\begin{array}{c} R \\ \diagdown \overset{+}{} \ \overset{-}{} \\ R'{-}N{-}O, \\ \diagup \\ R'' \end{array}$$ ammonium ions, $$\begin{array}{c} R \\ \diagdown \overset{+}{} \\ R'{-}N{-}R''', \\ \diagup \\ R'' \end{array}$$ and immonium ions,

$$\begin{array}{c} R \\ \diagdown \overset{+}{} \diagup \\ C{=}N \\ \diagup \diagdown \\ R' \end{array}$$

The nitrogen atom in its ground state is represented by the electronic structure $(1s)^2(2s)^2(2p_x)^1(2p_y)^1(2p_z)^1$. This distribution of electrons would predict the formation of three orthogonal σ bonds in amines, the lone pair not having any preferred orientation. These predictions are completely contrary to experimental results; in fact, the nitrogen atom has a tetrahedral structure, the electron pair occupying one of the corners of the tetrahedron.

In the case of ammonia, R=R'=R''=H, the intervalency angle is 107°; for trimethylamine, R=R'=R''=Me, the intervalency angle is 108° 7'. The structure of nitrogen in the amines can be understood by the hypothesis of sp³ hybridization of the atomic orbitals:

$$(1s)^2(2sp^3)^2(2sp^3)^1(2sp^3)^1(2sp^3)^1.$$

One of the hybrid orbitals is occupied by two electrons, that is the electron pair which acts as if it were a substituent. The nitrogen atom is iso-, electronic with the carbon atom of a carbanion (*see* page 16), and like the latter it is sterically labile. The inversion of the tetrahedral structure of an amine is usually very rapid at normal temperatures, as a result of passage through a planar intermediate in which the nitrogen atom possesses sp² hybridization.

Nuclear magnetic resonance, n.m.r., allows the stability of a structure to be estimated by studying the resonance of the protons of the molecule at varied temperatures. It has been shown, for dibenzylmethylamine, for example, that the frequency of inversion is of the order of $76 \sec^{-1}$ at $-146°C$. A complete inhibition of inversion has been observed only for certain exceptional amines; polycyclic amines in which the nitrogen atom is at a bridgehead (*see* page 100) and N-halogenated aziridines such as (A) (below):

(A)   (B)

In amine oxides $R(R')(R'')\overset{+}{N}-\overset{-}{O}$, amino-boranes $R(R')(R'')\overset{+}{N}-\overset{-}{B}H_3$ or ammonium ions $R(R')(R'')(R''')\overset{+}{N}$ the extra linkage is made with the aid of the lone pair, maintaining the tetrahedral structure around nitrogen. The configuration of the nitrogen atom is sterically stable; for example, trimethylamine oxide, represented by structure (B) (above).

Double bonds $\diagdown C=N-$ use an atom of nitrogen in the hybridization state sp²:

$$(1s)^2(2sp^2)^2(2sp^2)^1(2sp^2)^1(2p_z)^1.$$

One of the sp^2 orbitals is occupied by two electrons, a free doublet. A π linkage is formed by lateral overlap using the 2p electrons of nitrogen and carbon. The nitrogen atom is planar, and the intervalency angles are close to $120°$ (Figure 1.6).

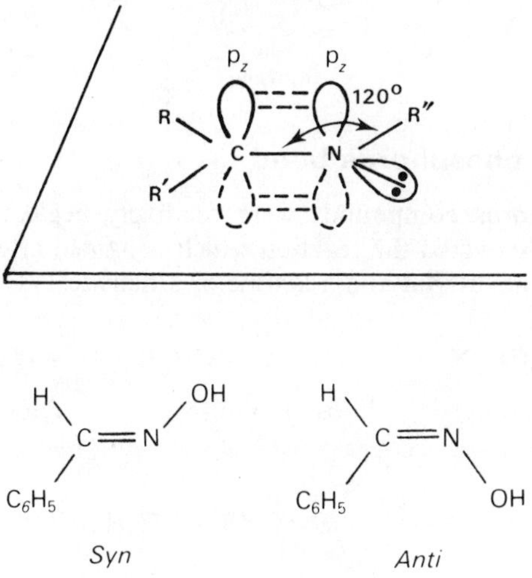

Isomeric benzaldoximes

Figure 1.6

The π bond imposes its planarity on the unsaturated system and makes geometrical isomerism possible. In the imines the activation energy for the reaction of isomerization is not usually very large and interconversion between geometrical isomers occurs rapidly, sometimes preventing the isolation of pure isomers at ordinary temperatures. On the other hand, in oximes many cases are known in which the geometrical isomers can be isolated. In Figure 1.6 the *syn* and *anti* isomers of benzaldoxime are shown. Here the description *syn* or *anti* refers to the relative position of H and OH.

The mechanism of *syn–anti* isomerization has recently been clarified by n.m.r. studies. Two mechanisms are possible: rotation about the C—M axis in a dipolar intermediate which is relatively rare; and inversion at the nitrogen atom which is the process that usually allows geometrical isomerization.

$$\begin{array}{c}R\\ \diagdown\\ \diagup\\ R'\end{array} C = N \begin{array}{c} \cdot\cdot\\ \diagdown\\ Y \end{array} \quad\longleftrightarrow\quad \begin{array}{c}R\\ \diagdown\\ \diagup\\ R'\end{array} \overset{+}{C} \rightarrow N \begin{array}{c} \cdot\cdot\\ \diagdown\\ \underset{\ominus}{Y} \end{array} \quad\rightleftharpoons\quad \begin{array}{c}R\\ \diagdown\\ \diagup\\ R'\end{array} C = N \begin{array}{c} \diagup Y\\ \cdot\cdot \end{array}$$

Rotation

$$\begin{array}{c}R\\ \diagdown\\ \diagup\\ R'\end{array} C = N \begin{array}{c} \cdot\cdot\\ \diagdown\\ Y \end{array} \quad\rightleftharpoons\quad \begin{array}{c}R\\ \diagdown\\ \diagup\\ R'\end{array} C = N - Y \quad\rightleftharpoons\quad \begin{array}{c}R\\ \diagdown\\ \diagup\\ R'\end{array} C = N \begin{array}{c} \diagup Y\\ \cdot\cdot \end{array}$$

Inversion

The carbon–phosphorus bond

Organophosphorus compounds were relatively neglected until 1954 when Wittig discovered the reaction which is named after him, and involves a phosphorus ylid and a ketone or aldehyde.

$$P(C_6H_5)_3 + RCH_2X \longrightarrow (C_6H_5)_3\overset{+}{P} - CH_2R\ \overset{-}{X} \xrightarrow[-HX]{base} (C_6H_5)_3\overset{+}{P}-\overset{-}{C}HR$$

Triphenylphosphine   Phosphonium salt   Phosphorus ylid

$$\text{Ylid} + \begin{array}{c}R'\\ \diagdown\\ \diagup\\ R''\end{array} C{=}O \longrightarrow \begin{array}{c}R'\\ \diagdown\\ \diagup\\ R''\end{array} C{=}CHR + (C_6H_5)_3\overset{+}{P}\ \overset{-}{O}$$

Triphenylphosphine oxide

Very many syntheses of olefins have been effected using this method. The numerous variations of the Wittig reaction have drawn the attention of organic chemists to derivatives of phosphorus, which have also proved in many cases to be interesting ligands in transition metal complexes.
Phosphorus $(Z = 15)$ possesses one electron fewer than sulphur:

$$(1s)^2(2s)^2(2p)^6(3s)^2(3p_x)^1(3p_y)^1(3p_z)^1.$$

The distribution in the M shell is exactly that in the L shell of nitrogen. The difference between phosphorus and nitrogen lies in the presence of d orbitals which are potentially available. Tricovalent compounds of phosphorus such as trimethylphosphine are formed from an sp^3 phosphorus atom:

$$(3sp^3)^2(3sp)^1(3sp^3)^1(3sp^3)^1,$$

the lone pair occupying a corner of the tetrahedron, as in the case of nitrogen. The phosphine oxides $(R)(R')(R'')\overset{+}{P}-\overset{-}{O}$ or the phosphonium ions

$(R)(R')\overset{+}{P}(R'')(R''')$ are also tetrahedral in structure, indicating sp^3 hybridization for phosphorus; the same is true for ylids. An ylid, $(C_6H_5)_3\overset{+}{P}$—CHR, may be considered as the reaction product of phosphine, $(C_6H_5)_3P$, with a carbene, :CHR, which possesses a vacant electronic orbital capable

of accepting the lone pair of the phosphine. The bond $\overset{\diagdown}{\underset{\diagup}{}}\overset{+}{P}-\overset{-}{C}\overset{\diagup}{\underset{\diagdown}{}}$ is written

$\overset{\diagdown}{\underset{\diagup}{}}P{=}C\overset{\diagup}{\underset{\diagdown}{}}$, the pair of electrons fixed on the carbon atom being accepted

by a 3d orbital of the phosphorus atom.

The double bond which results in this way makes use of a particular type of π bond, p_π—d_π. Double bond character is absent in nitrogen ylids,

$\overset{\diagdown}{\underset{\diagup}{}}\overset{+}{N}-\overset{-}{C}\overset{\diagup}{\underset{\diagdown}{}}$, because the L shell of the nitrogen atom does not contain

any d orbitals.

The organic compounds in which phosphorus is pentacovalent have been studied recently. In the molecules in which the phosphorus atom is pentacoordinate it is at the centre of a trigonal bipyramid, the substituents being at the corners.

There are three fluorine atoms at the base (equatorial fluorines) and two fluorine atoms (marked F*) on the three-fold axis (axial fluorines) PF₅ gives a single n.m.r. signal for fluorine at ordinary temperatures although there are two types of fluorine atom in the bipyramid. This experiment implies a rapid exchange mechanism between the equatorial and axial fluorine atoms. The mechanism of intramolecular exchange is still not well understood.

Intervalency angles

The magnitude of intervalency angles is in principle directly related to the nature of the atomic orbitals which form the interatomic linkages. Some typical values for intervalency angles, which reflect the hybridization of the orbitals used, were shown on page 11. The relationship between these angles and the nature of the atomic orbitals around an atom A can be summarized:

| $\overset{90°}{\underset{p}{A}}$ | $\overset{109°}{\underset{sp^3}{A}}$ | $\overset{120°}{\underset{sp^2}{A}}$ | $\overset{180°}{\underset{sp^1}{A}}$ |

In practice one often comes across deviations from these theoretical values. The reasons are either steric or electronic in origin; for example, the angular distribution of bonds around an atom of tetrahedral carbon (tetrasubstituted carbon). The molecule C*abcd* may be represented by placing the linkages C—*a* and C—*b* in the plane of the paper; the linkage C—*d* would be orientated toward the back and is indicated with a dotted line; the linkage C—*c* would lie toward the observer.

Intervalency angles around a
tetrahedral carbon atom

θ_{ab} θ_{ac} θ_{ad}
θ_{cd} θ_{bd} θ_{bc}

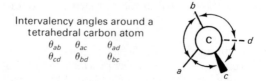

There are six intervalency angles. When the four groups *a*, *b*, *c* and *d* are identical, that is to say in structures Ca_4, the carbon atom is at the centre of a regular tetrahedron and all the angles θ_{aa} are equal to 109° 28′; for example, in CH_4 or CCl_4. In molecules where the groups are different, for example with Ca_3b, the symmetry of a regular tetrahedron is reduced. A certain flexibility of the intervalency angles is, indeed, possible, and a permanent deformation of the regular tetrahedron will result from the nonequivalence of the interactions *a*—*a* and *a*—*b*. The examination of various tetrasubstituted carbon atoms leads to an important conclusion. The ideal angle of 109° 28′ is found only rarely in organic molecules, because of angular deformations. The molecular models which are commercially available use tetrahedral carbon atoms as their main element. This element should only be used for the construction of symmetrical molecules of type Ca_4. In molecules which have lost the symmetry of a regular tetrahedron one cannot expect to build models which truly represent the molecular geometry. The determination of the structure of molecules by X-ray diffraction (in the crystalline state) or by electron diffraction or microwave spectroscopy (in the vapour phase) has given valuable information on interatomic distances and intervalency angles. If one takes methane as the point of reference one finds that the intervalency angle varies with the degree of substitution of the carbon atom:

The methylene $(-CH_2-)$ grouping of alkanes or of cyclohexane is characterized by an intervalency angle $\overset{\frown}{C \quad C}$ of the order of $112°$.

In molecules Ca_2b_2 in which the a groups are very large, steric repulsions force an increase in θ_{aa}. Thorpe and Ingold have postulated an effect, which bears their name, on the simultaneous diminution of θ_{bb}.

Thorpe-Ingold effect

The flexibility of intervalency angles makes possible the syntheses of substances of which the existence could never have been predicted on the basis of a regular tetrahedral carbon atom. In Figure 1.7 a number of these compounds are illustrated. Only cyclopropane will be discussed here. The angle C—C—C is $60°$, which makes it impossible to form a σ C—C bond by the axial overlap of sp^3 orbitals, as indicated on page 12. A partial overlap is nevertheless possible (Figure 1.7), but the atoms of the C—C bond are less strongly bonded than in normal bonds and the σ electrons are more delocalized.

This type of bond is sometimes called a bent bond or 'banana bond', because of its shape.

Cyclopropane has some electronic properties which show a certain analogy with those of a double bond. This is because of the electronic density in the plane of the ring, caused by the delocalization of the σ electrons.

An important and very general point can be made in connection with cyclopropane; there is not always a coincidence between the intervalency angle (or interatomic angle) and the interorbital angle. In the case of cyclopropane, if we presume that the three carbon atoms are in the sp^3 state of hybridization, the interorbital angle is $109° 28'$ and the angle between the bonds is $60°$. It should be emphasized that only the inter-

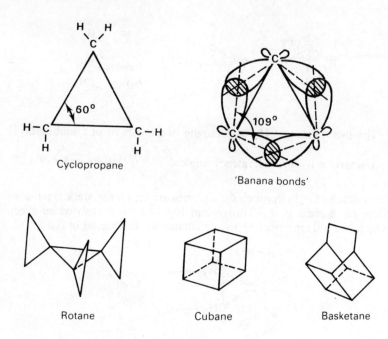

Cyclopropane

'Banana bonds'

Rotane

Cubane

Basketane

(Only C—C bonds are shown)

Figure 1.7 Examples of some strained saturated hydrocarbons that have been synthesized

valency angle is directly accessible by the methods of structural analysis, which give the relative positions of atoms. The interorbital angle is calculated from hypotheses on the nature of the bonds. The difference between the interorbital angle and the intervalency angle is an indication of the poor overlap of the atomic orbitals and of the instability of the bond. For cyclopropane $\Delta\theta = 109 - 60° = 49°$. As early as 1911 Baeyer observed the importance of the angular deformation $\Delta\theta$, which is the cause of angle strain. Angle strain (or Baeyer strain) is the energy which is needed to deform, by the value of $\Delta\theta$, the normal intervalency angle, assumed to be without strain. The angle strain T, in $kJ\,mol^{-1}$, is proportional to the square of the angular deformation $\Delta\theta$ in degrees when this is not too large:

$$T = K(\Delta\theta)^2$$

$K = 4.2 \times 10^{-2}$ in C—C—C chains. This formula shows that a deformation of the order of 10° for an angle C—C—C requires relatively little energy, about $4\,kJ\,mol^{-1}$. This is an economic means of reducing steric interactions within molecules.

In the trigonal carbon atom, the planarity of the three bonds and the 120° intervalency angles are a direct consequence of the sp² state of hybridization of the carbon atom. A deformation of these intervalency angles from the theoretical value of 120°, in the plane of the bonds, is possible. It results from the nature of the three groups which are attached to the trigonal carbon atom. In symmetrical structures of the type Ca_3 the intervalency angles are exactly 120°. The simplest example is the carbonate anion:

If the trigonal carbon atom is of the type Ca_2b or $Cabc$ the departures from 120° are predictable, and have been detected:

If an olefin has substituents with very large steric requirements, angular deformations in the plane of the double bond are sometimes insufficient for relieving steric strain. Ollis has recently shown that a reduction in steric interaction in (I) results from an out-of-plane deformation. The double bond is no longer planar and labile isomers (Ia) and (Ib) have been detected. They are the consequence of the 'folded' structure of the double bond.

| (I) | (I$_a$) | (I$_b$) |

These examples are very rare, and the principle of the planarity of a double bond in an ethylenic hydrocarbon can be considered as general. It is upon this principle that Bredt's rule depends. This rule, postulated in 1924, states that it is impossible to have a double bond at the head of a bridge in bridged polycyclic molecules.

Possible   Impossible

A more general statement of Bredt's rule is that a planar trigonal carbon atom cannot be placed at the head of a bridge, when this position requires a pyramidal atom, because of the geometry of the molecule.

Bredt's rule can be applied successfully to polycyclic systems containing small rings, which are already severely strained and cannot accept any increase in strain from the double bond at the bridgehead. It is, on the other hand, frequently broken when rings are five-membered or six-membered, despite the fact that Bredt established his rule initially with camphor:

Bredt's rule

An exception to Bredt's rule
Compound prepared in 1966
by Marshall and Wiseman

Length of bonds

Interatomic distance, or internuclear distance, is a function of the nature of the bonding atoms and the atomic orbitals which are used for forming the bond. Table 1.1 shows the length of some of the principal bonds encountered in organic chemistry, with the atomic orbitals which are used to form them. Notice that a double bond is shorter than a single bond.

Table 1.1   *Bond lengths* (nm)

C—C	0.154		C=C	0.133	C≡C	0.121
C—O	0.143		C=O (aldehyde, ketone)	0.121		
C—N (amine)	0.147	C=N	0.127	C≡N	0.115	
N—H	0.101		O—H	0.096	C—H	0.109	
C—F	0.135		C—Cl	0.177	C—Br	0.191	C—I	0.211

Values of bond lengths in Table 1.1 vary relatively little from one compound to another, because bonds are difficult to deform by extension or compression along the axis of a bond. The energy E_d (in $kJ\,mol^{-1}$) for the deformation of a bond is to a first approximation proportional to the square of the extension or compression (Δd) (in nanometers):

$$E_d = K(\Delta d)^2,$$

where $K = 1470$ for a C—C bond, and $K = 2940$ for a double bond C=C.

An extension of 0.005 nm in a C—C bond requires an energy of the order of 0.04 kJ mol^{-1} (variation in its length of 3%), whereas the same amount of energy is capable of effecting an angular deformation of more than 10% of the intervalency angle of a tetrahedral carbon atom.

A useful concept has been put forward, based upon a comparative examination of various bond lengths. In a bond A—B one assigns to each atom an atomic covalency radius. This fictitious radius r is such that the sum, $r_A + r_B$, shall be equal to the length d of the bond A—B. The value of the concept of the covalency radius of an atom lies in the fact that this quantity is independent of the nature of the atom to which it is bonded.

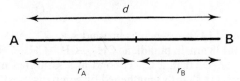

d = internuclear distance; r_A, r_B = covalency radii

For example, C—C=0.154 nm, from which one calculates $r_C = 0.077$ nm. C—H=0.109 nm, from which $r_H = 0.109 - 0.77 = 0.032$ nm. This value is found in virtually all A—H bonds. Table 1.2 shows the prin-

Table 1.2 *Atomic covalency radii* (nm)

⋝C—	0.077	>C=	0.0665	—C≡	0.0605		
H—	0.03						
—O—	0.074	O=	0.062				
>N—	0.074	—N=	0.062	N≡	0.055		
—S—	0.104	>P—	0.11	⋝Si—	0.117		
F—	0.064	Cl—	0.1	Br—	0.114	I—	0.133

cipal mean covalency radii. One can estimate the precision of calculations made with the aid of covalency radii, using the following molecule, of which the geometry has been experimentally determined:

$$
\begin{array}{c}
\text{H} \qquad\qquad \text{O} \\
\diagdown \qquad\quad \diagup\!\diagup \\
\text{H—C—C} \\
\diagup\quad |\ \ | \quad\diagdown \\
\text{H} \quad {}^1\ \ {}^2 \quad {}_3\text{C}\!\equiv\!\text{N}
\end{array}
$$

	Calculated	*Found*
C_1-H	$0.077+0.03=0.107$ nm	0.109 nm
C_1-C_2	$0.077\times2=0.154$ nm	0.149 nm
$C_2=O$	$0.066+0.062=0.128$ nm	0.122 nm
C_2-C_3	$0.077\times2=0.154$ nm	0.147 nm
$C_3\equiv N$	$0.06+0.055=0.115$ nm	0.116 nm

Influence of conjugation on molecular geometry

Systematic deviations from theory are evident in measurements of inter-atomic distances in systems in which double and single bonds alternate, or in which a double bond and an atom carrying a lone pair are linked.

For example, in amines $\overset{\diagdown}{\underset{\diagup}{-}}C-NH_2$ the mean value of the C—N bond

distance is about 0.147 nm, whereas in amides $R-\overset{\overset{\textstyle O}{\|}}{C}-NH_2$ the distance C—N is about 0.138 nm. In butadiene $CH_2=CH-CH=CH_2$ the central C—C bond is shorter than a C—C bond in a saturated molecule, and the double bonds are slightly lengthened by comparison with the double bond in a mono-olefin:

Structure of butadiene

The two double bonds of butadiene cannot be considered independently of each other. *Conjugation* between the two double bonds is responsible for the peculiar behaviour of butadiene (and is evident also in its chemical reactivity). It arises from the overlap of the two π electron clouds. If one considers the atomic p orbitals of four carbon atoms, quantum mechanics shows that it is possible to construct four molecular orbitals. In the ground state the four π electrons occupy in pairs the orbitals Ψ_1 and Ψ_2. Note that only the orbital Ψ_2 recalls the classical formula of butadiene, with strong electron density between C_1 and C_2, C_3 and C_4. The molecular orbital Ψ_1 is such that two π electrons are very delocalized.

In these conditions, the molecular orbital Ψ_1 confers a partial double bond character on the bond C_2-C_3, whereas C_1-C_2 and C_3-C_4 lose some of their double bond character in the overlap between the two π

clouds. The consequence is a shortening of the central bond and a lengthening of the double bonds.

The structure of butadiene may be discussed equally well in terms of *resonance*. This concept was introduced by Pauling in 1940. A number of formulae, each imperfect, are attributed to a compound (contributory structures). The true formula, which should explain the nature of all the bonds, is the weighted average of the various contributing structures. The double arrow ↔ indicates not an equilibrium but an 'overlap' between the formulae. In this very empirical treatment[1] one tries in a way to construct a photofit picture of the bonds in a molecule by superimposing a number of incomplete diagrams. In each contributing structure the atoms have the same position, only the disposition of the valence electrons changes.

$$CH_2 = CH - CH = CH_2 \longleftrightarrow \overset{-}{C}H_2 - CH = CH - \overset{+}{C}H_2$$
$$(2)$$

$$\longleftrightarrow CH_2 ==== CH ==== CH ==== CH_2 \longleftrightarrow \cdots$$
$$(3)$$

Resonance structures

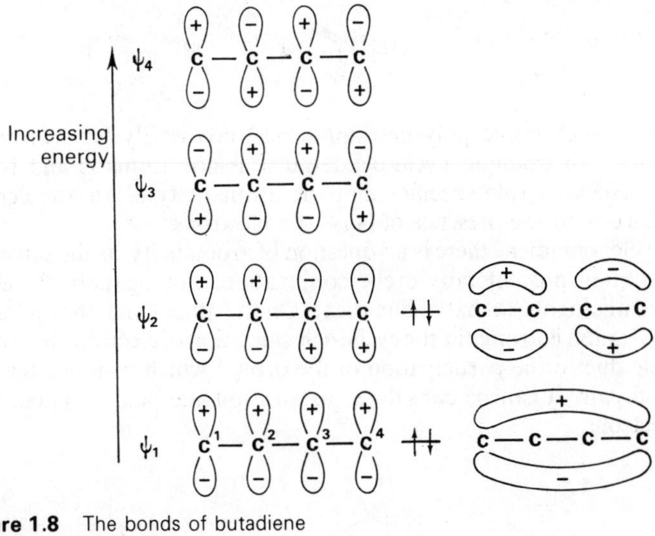

Figure 1.8   The bonds of butadiene

In Figure 1.8 several contributing structures or mesomers of butadiene are shown, formulae 2 and 3 giving some double bond character to the central bond of butadiene.

If one limits oneself to looking at the resonance 1↔3 one notices an

[1] The theory of resonance or mesomerism is now largely obsolete, but it is still widely used by organic chemists to express the electronic properties of a molecule in a formula.

analogy with the description of the π system by the molecular orbitals Ψ_1, Ψ_2. However, 3 corresponds to a delocalization of four π electrons among three bonds, in a structure having a certain 'weight', whereas two electrons only are distributed between the three C—C bonds in the Ψ_1 orbital.

The variation in the lengths of bonds is not the only consequence of conjugation. For conjugation to be effective a lateral overlap of the p atomic orbitals is also required, necessitating the coplanarity of the two double bonds. Butadiene and the amides are discussed on page 119.

It is possible for planar cyclic systems with alternating single and double bonds to show an enhanced conjugation which manifests itself by a set of properties (stability, susceptibility to electrophilic substitution, etc.) characteristic of aromaticity. Benzene is the best-known aromatic system. The molecule is planar, the carbon atoms are at the corners of a regular hexagon and the delocalization of the π electrons is complete. Each C—C bond has a length of 0.139 nm, a value intermediate between a double bond (0.133 nm) and a single bond (0.154 nm).

Not all such cyclic polyenes (annulenes) necessarily have aromatic character. For example, cyclo-octatetraene is not aromatic and is not planar. Hückel's rule specifies that the aromaticity of an annulene is associated with the presence of $(4n+2)$ π electrons.

In cyclopentadiene there is no question of aromaticity, as the saturated carbon atom prevents any cyclic conjugation. Conjugation, therefore, remains analogous to that of butadiene. On the other hand, the cyclopentadienyl anion is aromatic, the cyclic delocalization of electrons becoming possible due to the participation of the orbital which contains the free electron pair. It can be considered as an aromatic planar system with six electrons.

Cyclo-octatetraene
nonaromatic

Cyclopentadienyl anion
aromatic

The most familiar aromatic nucleus is the benzene ring. It is found in derivatives of benzene and in complex molecules such as naphthalenes, phenanthrenes, etc. The regular hexagonal structure is not always conserved if there are strong steric constraints in the molecule:

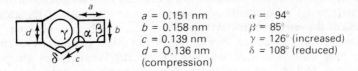

a = 0.151 nm	α = 94°
b = 0.158 nm	β = 85°
c = 0.139 nm	γ = 126° (increased)
d = 0.136 nm	δ = 108° (reduced)
(compression)	

Example of the deformation of the benzene nucleus

Van der Waals radii and nonbonded interactions

The approach of two atoms A and B, initially far apart, results in a variation in the potential energy of the system in consequence of interaction terms which depend upon the interatomic distance. The two atoms may bond to form a linkage; or they could equally remain nonbonded, whether forming part of the same molecule or of two different molecules. The forces of such nonbonded interactions are called respectively intramolecular or intermolecular. They are, in principle, the cause of steric interactions.

The curve describing the potential energy of a system of two atoms A and B as a function of their distance d is characteristic and shows a minimum (Figure 1.9).

The system reaches maximal stability at the distance d_0. An approach

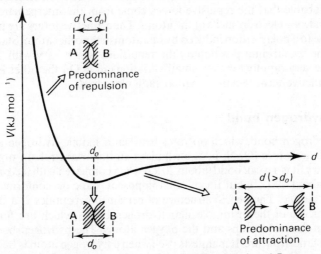

Figure 1.9 Potential energy of a system of two atoms, A and B, separated by distance d

of the atoms to a smaller distance than d_0 is energetically unfavourable due to the existence of repulsive forces, called *van der Waals forces*. If, on the other hand, one considers two atoms at a great distance from each other $(d > d_0)$, a potential curve shows that an approach of the atoms reduces the energy of the system. The forces which exist between the atoms are thus attractive; they are sometimes called *London forces*. They are proportional to $1/d^6$, whereas the van der Waals forces are proportional to $1/d^{12}$. At the distance d_0 the attractive and repulsive interaction terms of the van der Waals forces are equal. It is useful to assume that d_0 is equal to the sum of two distances characterizing each atom, called the van der Waals radii of the atoms. Two nonbonded atoms cannot come to closer distance than the sum of their van der Waals radii without an acquisition of energy to overcome the steric compression. The arrangement of molecules in a crystal depends largely on the van der Waals radii of the atoms which are in contact with each other. The principal van der Waals radii of the atoms most commonly encountered in organic chemistry are shown in Table 1.3. The van der Waals radii are always greater than the covalency radii by about 0.08 nm.

Table 1.3   *Van der Waals radii* (nm)

C	0.15	H	0.12	O	0.14	S	0.185	N	0.15
F	0.135	Cl	0.18	Br	0.195	I	0.215	P	0.19

The origin of van der Waals and London forces is not well understood. It is probable that the repulsive forces come from the interpenetration of the valency electron shells of the atoms. The attractive forces are probably due to a polarization induced by one atom in the other atom. Starting from the equilibrium position d_0 the repulsions which vary with $1/d^{12}$ increase very rapidly for even small variations in d. On the other hand, the attractive term remains small at radii close to d_0.

The hydrogen bond

The hydrogen bond, which operates between a mobile hydrogen atom and a heteroatom (O, N, F, S, etc.), plays an important role in organic chemistry. It is a weak bond (about 20 kJ) when compared with a covalent bond (about 400 kJ), but it can often impose a shape or 'conformation' on a molecule. The helical structure of certain polypeptides is a direct consequence of the intramolecular hydrogen bonds which are formed between the NH groups and the oxygen atoms. The hydrogen bond is electrostatic in nature. It manifests itself when a hydrogen atom is bonded to a very electronegative atom which confers on it a slightly positive character. The hydrogen bond is fairly directional; it is strongest when the three atoms A—H...B are collinear. An anion or the π doublet of a double bond can sometimes play the part of the electronegative B atom.

Conclusion

A number of fundamental parameters have been defined which are involved in determining the geometry of the molecule, and which are closely related to the nature of the bonds which join the atoms one to another.

| Elongation C—C | Opening of intervalency angles | Torsion (rotation) about C—C |

Different possibilities for minimizing interactions between atoms A and B

Nonbonded interactions between atoms are important not only in intermolecular movement, but also in giving to any one molecule its preferred shape by an interaction of deformations which has the effect of minimizing intramolecular strain. Molecular deformations result from variations in the bond lengths and the interatomic angles. Another way of changing the relative disposition of atoms, or groups of atoms, consists of rotations around bonds. This aspect will be discussed in the chapter on conformational analysis.

2

Molecular models, planar representation of molecules

For many years chemists did not attach great importance to the true shape of molecules. They were only interested in the planar formulae of chemical compounds, which contain useful information about the linkages of atoms with each other. With the development of modern chemistry it has, however, become indispensable to consider molecular geometry carefully in order to understand or to predict the physical or chemical properties of the compounds one is studying. The examination of molecular models is always advisable as it often allows one to understand:

(a) the privileged forms or conformations,
(b) the interatomic distances between nonbonded atoms,
(c) steric hindrance around a functional group, and
(d) the possibilities of stereoisomerism.

Principles of molecular models

Two types of model are used at present: skeletal models and space-filling models. Skeletal models give a general view of the molecular geometry, to a given scale. The intervalency angles and the distances between atoms (represented in the form of points or small balls) can easily be measured. In these models an interatomic bond A—B is represented either by a rod of a length equal to the sum $(r_A + r_B)$ of the covalency radii of the atoms (Figure 2.1), or by joining together two rods of which each represents one of the covalent radii.

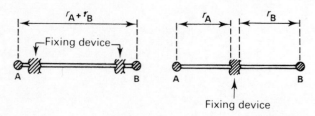

Figure 2.1 Method of forming an A—B bond

Each type of atom, A, B, etc., is constructed with a certain number of fixing devices, or of rods with such devices, so orientated as to maintain correct intervalency angles.

Space-filling models deal with the intervalency angles and bond lengths, of course, but they also introduce a new idea, the van der Waals radii of the atoms. Each atom is represented by a portion of a sphere having the van der Waals radius as its radius. The external form of the molecule is clearly seen by the observer, but the structure of the molecular skeleton is no longer perceptible. The jostling which results between two atoms which are forced into contact implies that they undergo van der Waals repulsion. The model obviously gives no quantitative indication of the magnitude of this interaction. In reality one can bring two nonbonded atoms much closer together than one would deduce from such a model, provided that a diminution in energy results for the molecular system as a whole.

The representation of orbitals is possible in some models; however, in those now commonly available this representation is very simplified.

Commercially available molecular models

Dreiding models
These are the most advanced skeletal models now available and are particularly well suited to research problems. Each sp^3 hybridized carbon atom is made of the junction of four metal rods, two hollow and two solid, arranged at 109° to each other (Figure 2.2). The C—C linkage is simply represented by pressing a solid rod into the hollow rod of the other atom. The junction is made on the principle of a press stud, which nevertheless does not prevent rotation around the C—C axis. When two atoms of sp^3 carbon are thus joined together their distance is exactly 3.85 cm, equivalent to 0.154 nm, as the scale used is 2.5 cm (1 inch) for 0.1 nm. A double bond $\diagdown$C$=$C$\diagup$ is represented by a rigid and planar system of five rods. Figure 2.2 shows, for example, —N, CO, —C≡C— and C_6H_5. Starting with a certain number of fundamental elements it is easy to construct a scale model of a molecule, the interatomic bonds taking the correct values as soon as the junctions between solid and hollow rods click into position. Dreiding models allow rotations around σ bonds, but make impossible the deformation of intervalency angles or the bending of bonds. To construct the strained rings in which the intervalency angles are abnormal one needs appropriate 'atoms'. For example, cyclobutane is formed starting with sp^3 carbon atoms of a particular type, having one intervalency angle of 90° and marked by a spot of colour. It is in practice possible to construct all types of molecules because of the very many structural elements which have been provided for in Dreiding models.

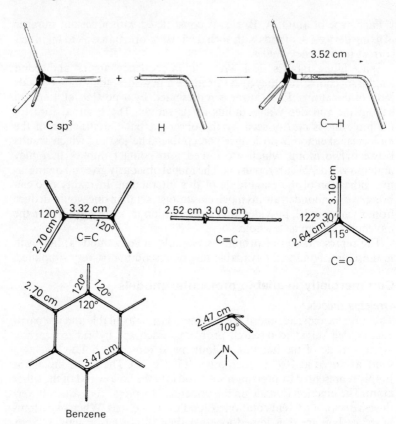

Figure 2.2 Some pieces from a box of Dreiding models

The excellent workmanship of these models makes it easy to measure angles and distances precisely. However, Dreiding models are not suitable for estimating steric hindrance or for forming an idea of the arrangement of orbitals in space.

Prentice–Hall models (framework molecular models)
These models are constructed of plastic tubes (representing bonds) which fix on to metallic rods signifying the valencies radiating from an atom.

Translator's footnote
One system of molecular models (available in the United Kingdom and elsewhere) that deserves mention is the Orbit/Minit series. These resemble the Prentice–Hall system, but have 'atoms' made from hard plastic, rather than metal. A wide range of atoms is supplied, including 1-, 2-, 3-, 4-, 5-, 6-, 8- and 12-coordinate arrangements, the 2- (and 3-) coordinate atoms being available with more than one set of intervalency angles. Strainless cyclopropane and cyclobutane rings can be constructed using the 12- and 8-coordinate atoms

Four small metallic rods, soldered in such a way as to maintain the 109° angle of tetrahedral carbon, represent sp^3 carbon atoms (Figure 2.3, (I)); sp^2 carbon atoms are made of three metallic rods, which are coplanar and make angles of 120° (II). Two rods perpendicular to the plane are available to represent the π bond. Octahedral complexes use the element (III). Atoms having intervalency angles of 90° can be represented by the basic elements (III) or (IV).

In principle, the elements (I) to (V) are adequate for the construction of all the molecules commonly encountered in organic chemistry. Thus an amine uses three of the four valencies of element (I), the last rod representing the lone pair of the nitrogen atom. The nitrogen atom of amides is planar, with intervalency angles of 120°, and therefore element (II) can be used.

The coloured plastic tubes are provided uncut; the user decides the scale of his own choice before cutting them. That suggested by the manufacturer is again 1 inch (about 2.5 cm) to 0.1 nm. The van der Waals radii of hydrogen are visualized in one direction (that of the CH bond) by means of a black and white plastic tube (*see* Figure 2.3).

The π cloud of double bonds is represented by a very simple arrangement which makes use of the existence of vertical rods in element (II). The orbitals of lone pairs of heteroatoms are equally well emphasized. For example, in formaldehyde (Figure 2.3) the π and n orbitals are apparent. The flexibility of the plastic tubes has both advantages and disadvantages. It is not possible to measure angles and interatomic distances with the same precision as with the Dreiding models. However, it is possible to construct very strained structures (such as cyclobutanes, *see* Figure 2.3) using tetrahedral atoms (I). The angle strain is absorbed by the curvature of the bonds, which thus represents the concept of the bent bond discussed on page 25). These models are inexpensive, and are very suitable for students while also being of great value in chemical laboratories.

SASM models
SASM models are designed to be used in either the space-filling or skeletal form. Each hydrogen atom is composed of a red hemisphere of which the radius is proportional to the van der Waals radius, and sp^3 carbon atoms are black spheres with four truncations, to which four atoms can be fixed. The dimensions of the atomic models are calculated in such a

respectively. In the Orbit set, 1 inch = 2.5 cm = 0.1 nm is the usual scale, whereas the Minit models work to about half this scale, and allow the construction of models of macromolecules of reasonable dimensions. As with the Prentice–Hall type, the main advantage is the ample supply of 'atoms' and 'bonds' that can be obtained at modest cost; however, every type of model has its own peculiar merits and disadvantages.

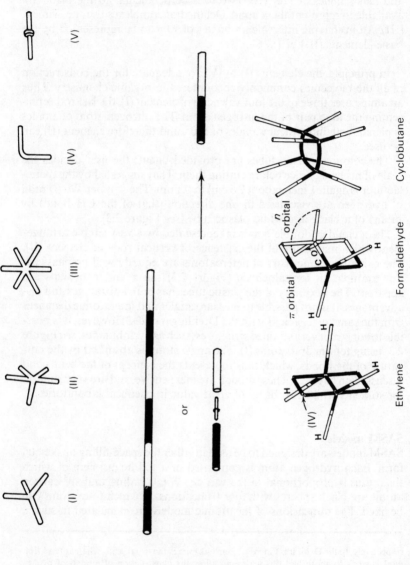

Ethylene

Formaldehyde

Cyclobutane

Figure 2.3   The constituents of a box of Prentice–Hall models

way that the covalent radii will be correct: the exterior volume marks the limit of the van der Waals radii (Figure 2.4).

All the atoms are provided with holes in which small rods can be fixed, or with press studs which allow the formation of bonds. The structures may be expanded by the use of longer metallic rods of which the dimensions are proportional to the bonds which are to be constructed. The spherical shells then lose their steric significance because of the change of scale and only symbolize the atoms.

The SASM models are precision-made on various scales for demonstrations in lectures or in the laboratory. Special models are also available for describing chemical bonds made by molecular orbitals, and in order to show the overlap between atomic orbitals.

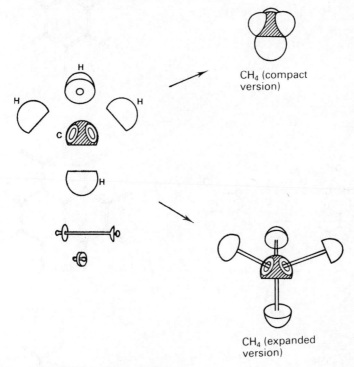

CH_4 (compact version)

CH_4 (expanded version)

Figure 2.4   SASM modecular models

Examples of problems studied with the aid of models

Three molecular models, available commercially, have been described, each having its advantages and its disadvantages. Many other systems are available, so that a variety of problems can be tackled. For example, the construction of a macromolecule is convenient if the model is light, the scale suitable and if a structural skeleton holds the macromolecule together.

The value of molecular models in solving some stereochemical problems is illustrated in Figure 2.5.

(A) Hydrogen bonds are ineffective beyond a certain distance (0.3 nm for the bond OH...O). The Dreiding model (2) of salicylaldehyde (1) arranged in its planar conformation shows that the distance O...H is

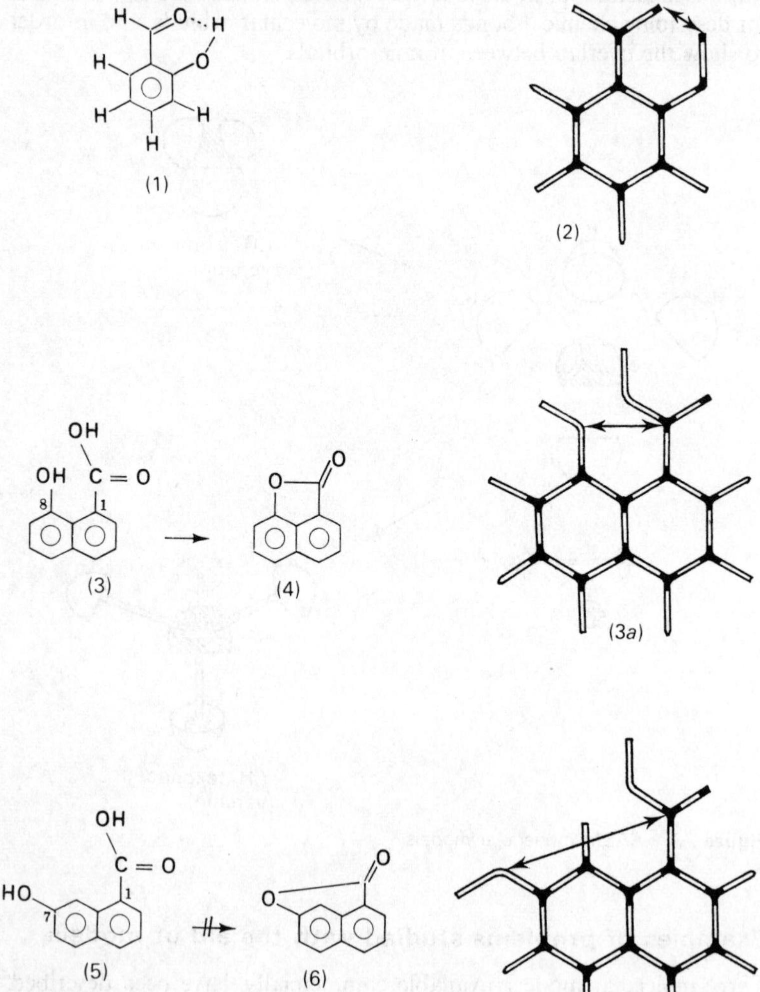

(1)

(2)

(3)   (4)

(3a)

(5)   (6)

Figure 2.5

(5a)

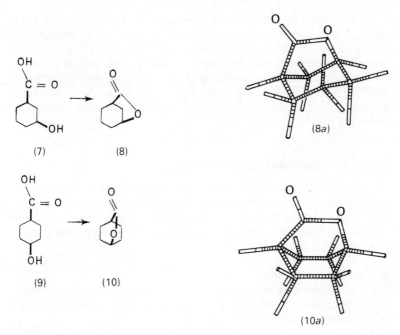

(7)   (8)   (8a)

(9)   (10)   (10a)

Figure 2.5—contd.

less than 0.2 nm, which is therefore compatible with the establishment of an intramolecular hydrogen bond. This is confirmed by experiment.

(B) The lactonization of a hydroxyacid only occurs when the molecular structure allows a suitable close approach of the alcoholic hydroxyl and

the carboxyl residue. In a lactone $\overset{\frown}{\diagup} O - C \diagup$ the distance O—C is of
$$\underset{O}{\overset{\|}{}}$$

the order of 0.135 nm. Examination of the molecular models of hydroxy-acids enables one to predict easily those that should lactonize. Thus 8-hydroxynaphthoic acid (3) easily gives the lactone (4), whereas the acid (5) is incapable of lactonization. Dreiding models (3a) and (5a) allow

measurement of the distance $\overset{H}{\diagdown}\quad\overset{OH}{|}$ In (3a) it is 0.29 nm
$$O \ldots C = O$$
$$\diagup \qquad \diagdown$$

and in (5a) 0.4 nm. The latter value is completely incompatible with the length of 0.135 nm for the O—C bond of a lactone. A considerable deformation of the molecular skeleton of (5) would be needed to allow the two functional groups to come so close together. The formation of the lactone is thus impossible.

(C) The shape or conformation of a molecule is often shown without ambiguity by constructing the molecular model. Let us consider, for example, the two isomeric hydroxyacids (7) and (9). These two compounds can be converted into the lactones (8) and (10), respectively. Molecular models (Prentice–Hall) (8a) and (10a) show that the cyclohexane ring is in the chair form in the lactone (8), but in the boat form in the lactone (10). These molecules are impossible to construct in any other conformations.

Ways of representing molecules

It is important to be able to draw a molecule in such a way that the essential features of the molecular architecture can be seen without always having to resort to the use of molecular models. The techniques used are diverse, and they make use of a number of conventions which it is essential to know.

The perspective drawing
This representation does not require any explanation as the compound is simply drawn as it would appear, once one has decided the direction from which it is observed (Figure 2.6).

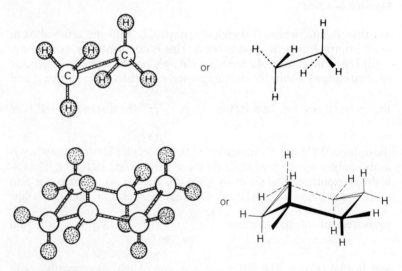

Figure 2.6

To give a certain sense of perspective to the drawing, bonds close to the observer are often shown with strong lines——— or◀▬▬ and bonds

placed towards the back of the diagram are represented by - - - - or ◁══.
Bonds placed in the plane of the page are drawn at a normal thickness.

Stereograms give a three-dimensional view of a molecular structure. They are useful for expressing the results of crystallographic study. Two views, corresponding to two observations at a slightly different angle, are placed side by side. A sense of perspective is obtained if each picture is viewed exclusively by one eye.

Newman projection

The molecule is observed along one bond. Two atoms are, therefore, projected one on top of the other (Figure 2.7). See also pp. 50, 55.

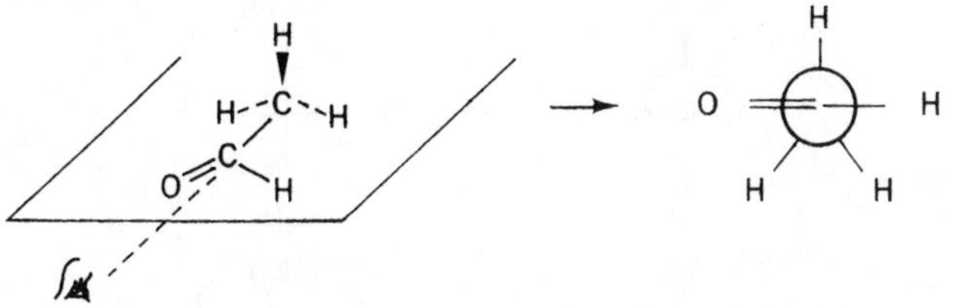

Figure 2.7 Newman projection

Fischer projection

In the perspective representation, the molecule is observed from the angle at which it is easiest to draw but there is no rule. The Newman projection, however, requires one to observe the molecule along the axis of the bond of which one wishes to clarify the stereochemical properties. The aim of the Fischer projection is to give information on the spatial distribution of the substituents around a tetrahedral carbon atom carrying different groups (asymmetric carbon atom). This question is discussed on page 109, in connection with absolute configuration. At present the principle is only considered using the example of lactic acid (11) (Figure 2.8). The central carbon atom is placed in such a manner that the CH_3-C-CO_2H bonds appear to be vertically aligned while the $H-C-OH$ bonds are shown horizontally. The essential convention which allows one to simplify the projection (11*a*) (in which the orientation of the bonds is shown by reference to the plane of the page) consists of always observing the carbon atom in such a way as to have the horizontal bonds directed towards the observer. In these conditions the simplified projection (11*b*) is sufficiently explicit and is equivalent to the model (11). The rules which are normally used for choosing the vertical chain and its orientation will be discussed on page 109. It should be remembered that a Fischer projection shows the distribution of groups about a carbon atom, and thus enables a corresponding molecular model to be constructed (for example, to go from (11*b*) to (11)). The Fischer projection can be used for any type of tetrahedral carbon atom, for example phenylacetic acid, $C_6H_5CH_2CO_2H$ (12), can be represented by (12*a*).

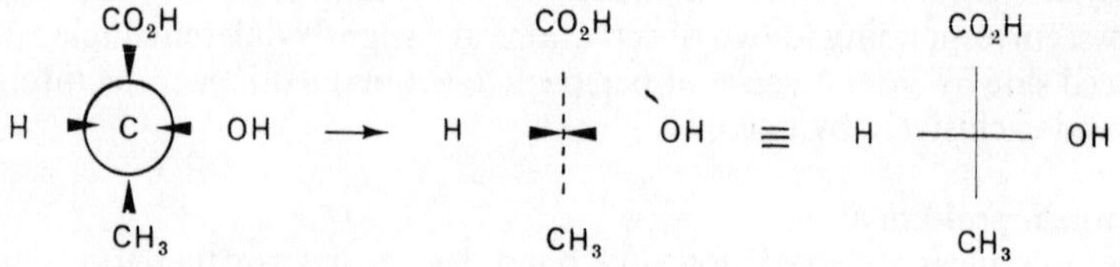

(11) (11*a*) (11*b*)

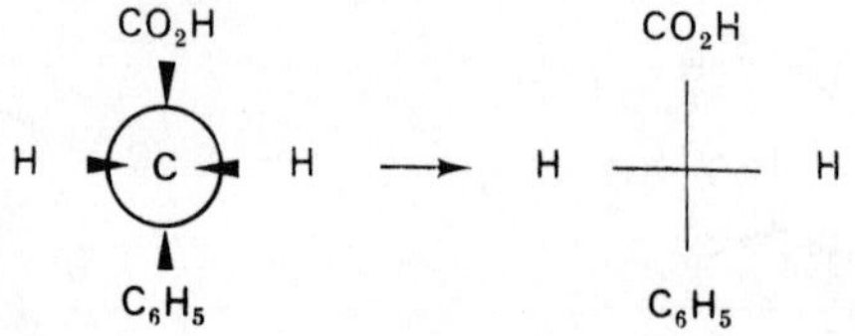

(12) (12*a*)

Figure 2.8 Fischer projection

Various representations and projections

The plane of the paper is a convenient plane for projecting a molecule, and the Fischer projection is a good example of using it. One avoids the need for any convention in projection, which is at the will of the chemist, providing that bonds in front of the plane are represented with strong lines and bonds at the back with dotted lines. Thus, lactic acid (11), observed according to diagram (13), can be represented by projection (13*a*) (Figure 2.9).

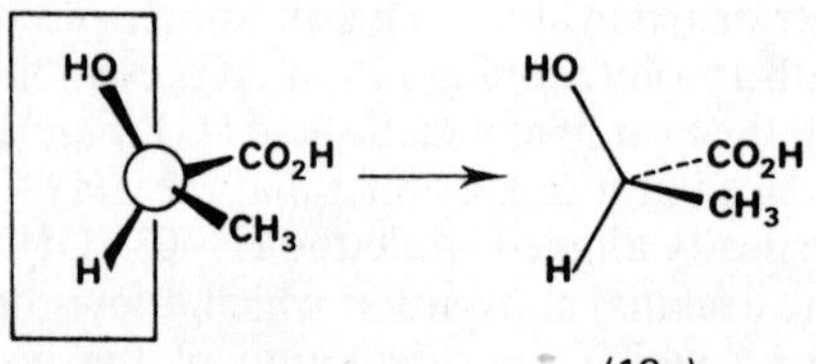

(13) (13*a*)

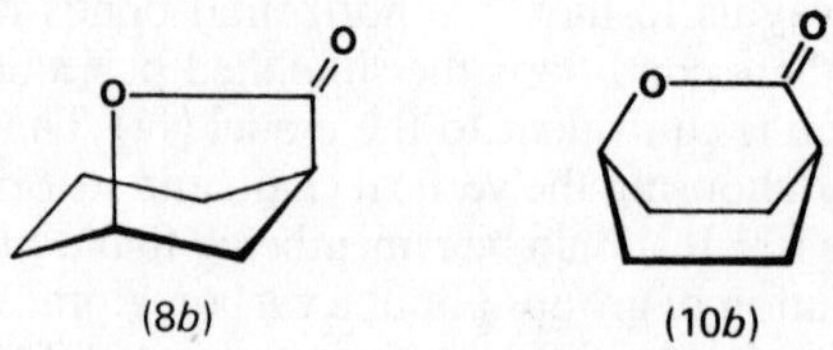

(8*b*) (10*b*)

Figure 2.9

A common convention with cyclic molecules involves arbitrarily representing the rings (whatever their true form) as planar. The orientation of the substituents can then be defined by reference to this fictional frame by the use of heavy or dotted lines. The planar formula is thus completed by stereochemical information, which however gives no indication of conformation. This convention has been used (Figure 2.5) for depicting the acids (7) and (9) and the lactones (8) and (10). The conformations (8a) and (10a) represent the true forms of the lactones (8) and (10), and can be drawn in a simplified manner by the perspective views (8b) and (10b) (Figure 2.9).

3

Conformational analysis

Starting with a structural formula, it is possible to predict the number of distinct molecular structures or *stereoisomers* which are theoretically possible, using rules which will be dealt with in Chapter 4. A planar formula always contains information (enabling the prediction of stereoisomerism) which every chemist must understand.

In the case of a planar formula to which there are no corresponding stereoisomers, the formula describes the compound in an unequivocal manner, for example:

$$CH_4 \text{ describes methane,}$$
$$CH_2{=}CH_2 \text{ describes ethylene.}$$

In these two examples a simple examination of the formula gives also the shape of the molecule. The carbon atom of CH_4 possesses sp^3 hybridization and the carbon atoms of ethylene the sp^2 hybridization, for which the structures are

Methane   Ethylene

Knowledge of the state of hybridization of carbon often helps to understand the molecular structures in three dimensions, without the use of molecular models. For example, in a cumulene, given that the digonal carbon atoms are of the type sp, it is possible to predict whether the molecule is planar or not (Figure 3.1).

In considering the above examples one should not assume that it is always easy to reconstruct the true shape of the molecule from its planar formula.

Consider the formula $CH_3{-}CH_3$ of ethane, the higher homologue of methane. The attachment of two tetrahedral carbon atoms does not deter-

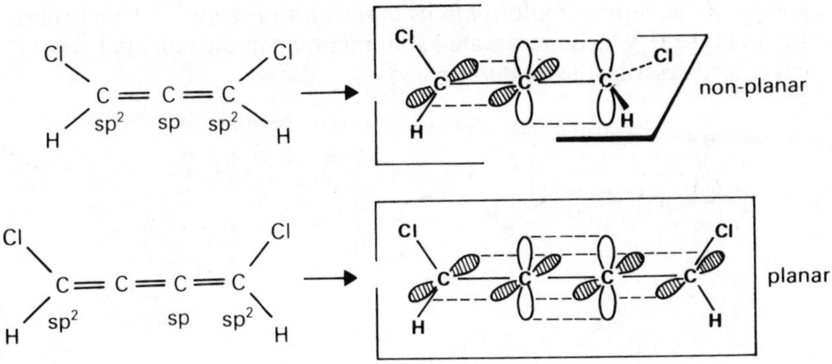

Figure 3.1

mine, unequivocally, the relative position of the three atoms of hydrogen situated on one carbon atom with reference to the hydrogen atoms on the other carbon atom. It has been known since the work of Pitzer, in 1936, that rotation is not completely free around the σ C—C bond. There is a barrier to the rotation when two C—H bonds are brought opposite one another. This hindrance does not prevent rotation, but it destabilizes the arrangement which is then described as 'eclipsed'.

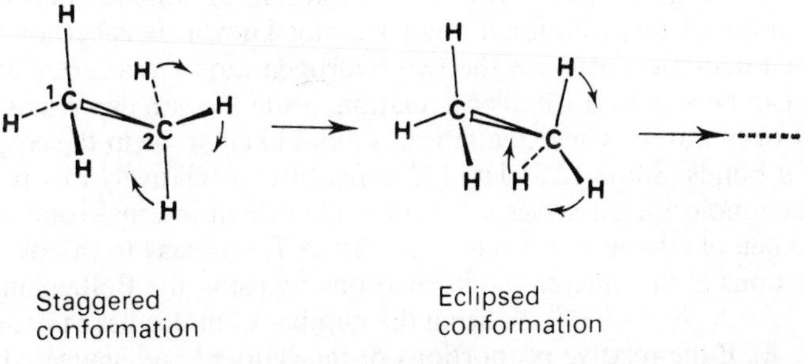

Staggered
conformation

Eclipsed
conformation

There is an infinity of possible positions for the hydrogen atoms situated on carbon atom 2, relative to those on carbon atom 1. Each arrangement is a conformation. The aim of conformational analysis is to examine the conformations adopted by molecules, and emphasize those which are energetically favoured. The representation of the conformations of ethane is helped by the use of the Newman projection (Figure 3.2), the molecule being observed along the C—C bond. If one regards one of the two atoms as fixed, a rotation of the second carbon atom continued through 360° brings each hydrogen atom back to its initial position. The eclipsed conformations (I) and the staggered conformations (II) are the characteristic arrangements encountered in the course of the rotation. The relative energy of the various conformations is shown in Figure 3.2. Conformations of type II, which are situated at a minimum in the potential energy curve, are described as *conformers*.

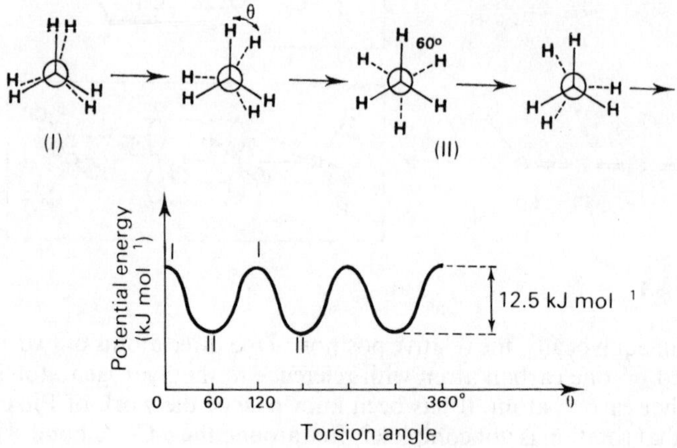

(I) → (II)

12.5 kJ mol^{-1}

Potential energy (kJ mol^{-1})

0 60 120 360° 0

Torsion angle

Figure 3.2

The eclipsed conformation (I) represents the rotational barrier for ethane. This energy barrier is called Pitzer strain, or torsion energy. The exact origin of this rotational barrier is not known. It is certainly not a steric interaction between the two hydrogen atoms that are eclipsed, which can be seen by a simple calculation, using the van der Waals radii of hydrogen atoms. The explanation is more likely to lie in the eclipsing of the σ bonds. Some 12.5 kJ mol^{-1} constitutes a relatively low barrier, and each molecule possesses a considerable rate of internal rotation. In a specimen of ethane at a given temperature T, it is easy to calculate the populations of the different conformations by using the Boltzmann formula $N_2 = N_1 e^{-(E_2 - E_1)/RT}$, N_i being the number of molecules i possessing energy E_i. If the relative proportions of the eclipsed and staggered conformations (I) and (II) respectively are calculated, one takes $E_2 - E_1 = 12.5$ kJ mol^{-1}, and finds (for $T = 25°C$) that there is only one molecule of eclipsed ethane for each one hundred and sixty of staggered ethane (II), that is to say a negligible proportion.

Torsional curve for propane

Propane, $CH_3CH_2CH_3$, gives a curve similar to that of ethane with a slightly higher barrier, 14 kJ mol^{-1}. The small difference between the rotational barriers of ethane (Figure 3.2) and propane shows clearly that the torsional energy does not originate in steric effects. The eclipsing of

CH with CCH_3 in propane is hardly more unfavourable than the eclipsing of CH with CH in ethane, despite the steric hindrance of the methyl group.

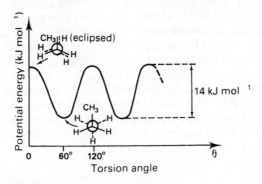

Torsional curve for butane

Butane, $CH_3CH_2CH_2CH_3$, possesses many characteristic conformations, which are shown in the Newman projection in Figure 3.3. The torsional curve shows two rotational barriers, the eclipsed forms (1) and

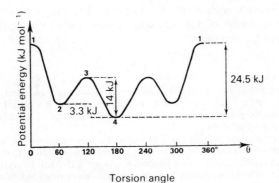

Figure 3.3

(3). The conformation (4), *anti* or extended, represents the potential mini-mum, starting from which the energies of the other conformations are measured. The lower barrier of butane, passing through conformation (3), is equivalent to that of ethane. Essentially the energy of pure torsion constitutes this barrier. However, in the higher barrier of butane, the eclipsing of two methyl groups in (1) introduces steric interactions, superimposed on the torsional energy of ethane, increasing the height of that barrier. The comparison of the energies of conformers (4) (*anti* or extended) and (2) (*gauche*) shows the latter to be destabilized by about 3.3 kJ mol^{-1}. This value mainly represents the steric interaction energy of two methyl groups at a dihedral angle of 60°, by comparison with the *anti* position. This result may be simplified by stating that the *gauche* interaction in butane is about 3.3 kJ mol^{-1}.

Gauche interaction
in butane

$+ 3.3$ kJ

(4)   (2)

This fundamental interaction is encountered in many organic structures. To summarize, butane exists in two preferred conformations, *anti* (4) and *gauche* (2). At room temperature the energy difference of 3.3 kJ mol^{-1} corresponds to about 1 mole of butane in the *gauche* arrangement for every 2 moles of butane in the *anti* arrangement.

Knowing the greater stability of the *anti* conformation, it is possible to predict the most stable conformation taken by a long chain, $-(CH_2)_n-$. In fact, we find the partial structure of butane if we con-sider successively each single C—C bond. The favoured partial con-formation is always of the *anti* type, which leads to a zig-zag planar arrangement for the chain:

Translator's footnote

A direct calculation of the equilibrium coefficient corresponding to an enthalpy difference of 3.3 kJ mol^{-1} gives at 25° a value of 4, not 2; however, there is a statistical factor of 2 which favours the *gauche* form, leading to an entropy advantage of R ln 2 for the latter, and thus to the equilibrium ratio of 2.

At room temperature roughly one kink occurs for every three or four C—C bonds in a long-chain compound $X[CH_2]_nY$, in the gaseous or dissolved state. Certain multiple *gauche* arrangements involve severe hindrance, and are forbidden. In crystals, when *n* is moderate (< 100), regular zig-zag conformations are observed, leading to an odd–even alternation in physical properties as the chain is extended, the $X—CH_2$ and $CH_2—Y$ bonds being alternately at about 112° and 180° to each other in the crystal.

Conformation of some acyclic unsaturated molecules

The hindrance to rotation round a single bond has also been observed in many unsaturated structures. Recently, microwave spectroscopy and nuclear magnetic resonance studies have given information about the height of rotational barriers. For example, acetaldehyde exists mainly in the conformation in which the hydrogen atom is eclipsed by the carbonyl group. In the preferred conformation of propionaldehyde it is the methyl group which eclipses the carbonyl group. The unforced eclipsing of a double bond by a single C—R bond seems to be a general phenomenon whenever R does not cause strong steric hindrance, for example when R=H or CH$_3$. The origin of this phenomenon, which is found also in enol ethers, still remains obscure. Many esters, for example methyl acetate, exist in a planar conformation in which the carbonyl group eclipses the substituent carried by the oxygen atom, for example CH$_3$:

| Favoured conformation: | R = H, acetaldehyde R = CH$_3$, propionaldehyde | Propene | Enol ether | Methyl acetate |

Butadiene exists at room temperature in two conformations, *s-cis* or *cisoid* and *s-trans* or *transoid*, in rapid equilibrium. The equilibrium strongly favours the *s-trans* isomer, nevertheless during the reactions of the Diels–Alder type equilibrium is displaced without difficulty towards *s-cis* butadiene, which is the only form able to effect the desired condensation:

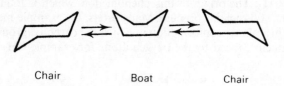

$$-\Delta G^\circ = 8 \text{ kJ mol}^{-1}$$

s-cis   s-trans

In butadiene the stabilization of the *s-cis* and *s-trans* conformations is due to the overlap of the two π orbital systems (conjugation). The rotational barrier is of the order of 21 kJ mol^{-1}.

Cyclohexane

Cyclohexane has played a fundamental role in the development of conformational analysis. The idea of nonplanar cyclohexane was first suggested by Sachse in 1890. The valence angles would be 120° in the planar structure, which would imply a considerable angular deformation. Sachse drew attention to the fact that certain conformations of cyclohexane, the chair conformation and the boat conformation, are without angle strain.

Chair   Boat   Chair

Mohr took up these ideas again in 1920, but without achieving general acceptance. Until 1950, chemists, apart from a few exceptions like the Dutch chemist Boeseken, saw no need to theorize about the conformations of cyclohexane, since the conversion of one chair into its inverse left the ring planar on average. Barton (Nobel Prize, 1969), using the work of the physical chemist Hassel (Nobel Prize, 1969), laid the foundations for the conformational analysis of cyclohexane. Hassel's experiments with various cyclohexanes, using X-ray diffraction and electron diffraction, showed that the stable form of cyclohexane is the chair form. If one examines the chair form in detail one sees two types of bond (other than those of the ring): the *axial* bonds directed perpendicular to the plane of the ring; the *equatorial* bonds slightly inclined to the plane of the ring.

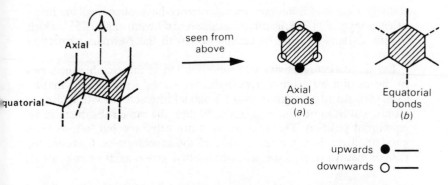

upwards ● ——
downwards ○ ——

Each carbon atom carries two hydrogen atoms, one axial (*a*) and the other equatorial (*e*). The hydrogen atoms having axial character can be separated into two groups. Three axial bonds are placed 1,3 to one another above the plane of the ring, and the other three axial bonds are similarly *cis* to one another and are orientated below the plane. The equatorial substituents are governed by the same rules; however, the relationships, *cis* or *trans*, between the equatorial groups are less evident, as each equatorial bond seems close to the plane of the ring. It is easy to avoid errors by considering the axial bonds simultaneously. For example, two equatorial hydrogen atoms arranged 1,3 to one another are orientated on the same side of the plane of the ring (relative stereochemistry *cis*); the two axial bonds at these corners are also *cis* but on the opposite face. Two equatorial hydrogen atoms at adjacent corners are *trans* to one another, as are the two axial hydrogens on these atoms. The representation of cyclohexane in the Newman projection along a C—C bond brings out various characteristic dihedral angles. All the bonds are staggered, which partly explains the great stability of cyclohexane.

Newman projection
of cyclohexane

If the value of 109° 24′ is assumed for the intervalency angle of carbon atoms, the dihedral angles of the ring will be 60°, with alternation of sign. The dihedral angles *ee* and *ea* are also 60°. Structural studies, however, have shown that the intervalency angle of carbon depends on its degree of substitution. It changes from 109° 24′ in methane to 112° in normal alkanes (*see* page 25). The latter value should be used for cyclohexane;

actually electron diffraction measurements have suggested an inter-valency angle of $111° 5'$, leading to an internal dihedral angle of $55°$, which slightly flattens the ring by comparison with the Newman projection shown above.

Barton related reactivity and conformation in the cyclohexane series by means of a very simple hypothesis. He started with the idea that an axial substituent, affected by two 1,3 diaxial interactions with hydrogen atoms, will be more hindered sterically than the same substituent in an equatorial position. This hypothesis is not arbitrary, but follows from good analogies between the cyclic and the acyclic series. Consider the case of methylcyclohexane, with the methyl group axial or equatorial:

(I)   (II)

In (I) one can see the structural element of *gauche* butane twice (for example by using a Newman projection along carbon atoms 1 and 2, with methyl and C_3 playing the roles of the methyl groups of butane). In II one observes two arrangements of the *anti*-butane type. It can thus be predicted that the conformation (I), with the methyl group axial, will be less stable by $2 \times 3.3 = 6.6 \text{ kJ mol}^{-1}$, relative to conformation (II), in which the methyl group is equatorial. Relative to (II), (I) shows two *gauche*-butane interactions. Equally, one can relate the instability of con-formation (I) to two 1,3 diaxial interactions methyl/hydrogen, which dis-appear when the methyl group becomes equatorial. The equilibrium (I)⇌(II) has been measured directly by spectroscopic methods, and in-directly. It is characterized by a difference of free energy:

$$-\Delta G^{\theta}_{\text{(II)-(I)}} = 7.5 \text{ kJ mol}^{-1}$$

The value $-\Delta G^{\theta} = 7.5 \text{ kJ mol}^{-1}$ is called the conformational preference for an equatorial position by methyl (or the A value of methyl, according to terminology proposed by Winstein). The larger a substituent (X) is, the less it tends to take up an axial conformation and the larger the A values becomes. The chair⇌chair conformational equilibrium is dis-placed towards the conformation in which X is equatorial. This aspect of the conformational analysis of cyclohexane will be dealt with in more detail later.

The problem of reactivity in the cyclohexane series can be approached by considering cyclohexanes rendered rigid, for example as a result of fusion with other rings. A reaction sensitive to steric hindrance will be slowed when the reaction site is axial. For example, the hydrolysis of acetates:

Slow   Rapid

Conversely, reactions accelerated by steric hindrance are facilitated by an axial reaction site. Axial hydroxyl groups are oxidized more rapidly than equatorial hydroxyl groups when the oxidizing agent is chromic acid. The relationships between conformation and reactivity will be discussed in the chapter on dynamic stereochemistry.

Inversion of cyclohexane; nonchair conformations

The deformations which transform one chair into the inverse chair (*see* page 58) necessarily require the ring to pass through a set of conformations of very different energy. The chair conformation represents the minimum energy. The conformation of maximum energy is called the inversion barrier, and it is measured, as in the rotation of acyclic compounds, by reference to the minimum energy. The inversion barrier (about $43 \, kJ \, mol^{-1}$) is believed to consist of a conformation described as an 'envelope', in which five carbon atoms are coplanar. This conformation can deform further to the boat form. If one constructs a molecular model of cyclohexane, for example with Dreiding models, it is easy to see the relative rigidity of the chair form and the great flexibility of the boat form. The cyclohexane boat form is deformed without difficulty giving a set of conformations called 'flexible'. The most stable of the flexible conformation is that described as the 'twist' conformation (in French *croisée*) which possesses three axes of symmetry of order 2. The twist conformation is often represented as viewed by the observer along one or other of the symmetry axes:

Envelope   Twist conformation

Different flexible conformations are, as it were, imprisoned between two energy barriers which hinder their easy transformation into the chair form. The easy interconversion between flexible conformations is called pseudorotation. Figure 3.4 shows the variations in potential energy of a cyclohexane molecule during the interconversion of one chair conformation to its inverse. This diagram shows some resemblance to that shown on page 51 for rotation around the central linkage of butane.

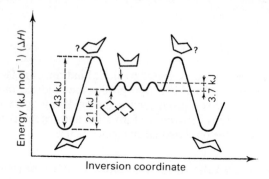

Figure 3.4

Translator's footnote
Although the shapes of stable molecules are often known accurately, the conformations of transition states remain a subject of debate. For example, the transition state between the chair and the flexible form of cyclohexane may have four adjacent methylene groups coplanar, with the remaining groups being one above and one below the plane so defined. It would thus resemble the cyclohexene molecule (p. 68), and would be directly intermediate between the chair and twist forms, both energy minima, whereas the envelope conformation represents an energy maximum and is intermediate between the chair and the boat conformations.

   The conformational analysis of flexible cyclohexane is of interest. In the twist form (D_2), three types of C—H linkage can be seen, for which the names twist-equatorial (*teq*), twist-axial (*tax*), and isochiral (*iso*) have been proposed; the latter term refers to the equal angles that each of the germinal C—H bonds make to the C_2 axis passing through two of the carbon atoms of a twist cyclohexane ring. In the boat form (C_{2v}), a transition state in the pseudorotation itinerary which interconverts the three equivalent twist-forms, the substituents can be called boat-equatorial (*beq*), boat-axial (*bax*), linear (*lin*) and perpendicular (*perp*). The two latter bonds originate from the unique pair of carbon atoms which lie on one of the mirror-planes of the C_{2v} form. It should be remembered that the energy difference between the twist and boat forms in cyclohexane itself is not large, and in more complex compounds the boat form, or conformations intermediate between boat and twist, may be more stable than the twist form.

Twist cyclohexane   Boat cyclohexane

The boat conformation is a little less stable than the twist conformation because of the severe interactions which exist between the two hydrogen atoms fixed on the 'masts'. Eclipsed interactions also exist between hydrogen atoms placed on neighbouring carbon atoms (see Newman projection), which introduce torsional energies analogous to those of ethane. The slight deformation which transforms a boat conformation into a twist conformation diminishes these interactions and the torsional strain.

Evidence for the inversion of cyclohexane: conformational isomerism

A sufficiently large reduction of temperature makes the molar fraction of the molecules of cyclohexane that are sufficiently 'hot' to be able to invert negligible, and literally freezes the chair conformation. Spectrometric methods can demonstrate the inversion and its freezing. Nuclear magnetic resonance (n.m.r.) shows a sharp signal for the twelve cyclohexane protons (Figure 3.5). This signal corresponds to an average spectrum; an axial proton, characterized by its coupling constants and its chemical shift, transforms itself into an equatorial proton by an inverstion of the ring, rapid on the timescale of n.m.r. The lowering of temperature widens the signal, which begins to split at a temperature called the temperature of coalescence T_c. By continuing to lower the temperature one can completely freeze out the conformational inversion, the six axial protons now appearing at $8.9\,\tau$ and the six equatorial protons at $8.4\,\tau$.

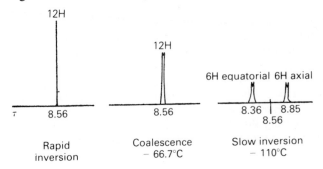

Figure 3.5   N.m.r. study of cyclohexane

A simple equation, the Gutowsky—Holm equation, relates the rate of conformational inversion k (number of inversions per second) to the coalescence temperature:

$$k = \frac{\pi \Delta v}{\sqrt{2}} \, s^{-1}$$

Δv being the gap (in ppm) between the chemical shifts of the two signals in the frozen conformation. For cyclohexane $k = 105\,\mathrm{s}^{-1}$ at $-66.7°C$.

It is useful to measure the height of an inversion barrier, as this enables one to predict at what temperature a labile conformation can be isolated. N.m.r. is a convenient method for calculating an inversion barrier, $\Delta G^{\pm}$.

Eyring has established a relationship, named after him, which relates the velocity of interconversion k, the temperature of coalescence T_c and $\Delta G^{\pm}$:

$$k = \frac{K'K_BT_c}{h} e^{-\frac{\Delta G^{\ddagger\pm}}{RT}}$$

The study of the n.m.r. spectrum of cyclohexane as a function of temperature leads to the following values for the parameters of activation for conformational inversion, at $-66.7°C$: $\Delta G^{\pm}=43\,kJ$, $\Delta H^{\pm}=38\,kJ$, $\Delta S^{\pm}=12\,JK^{-1}$. The possibility of isolating a conformation by working at a suitable temperature has highlighted a new type of isomerism, conformational isomerism. In 1966, Jensen studied the n.m.r. spectra of monochlorocyclohexane at various temperatures. The coalescence temperature T_c is observed at $-115°C$. Going down to $-150°C$ the conformational inversion is completely stopped. In contrast to the case of cyclohexane, the two conformations are not equivalent; each is characterized by its own n.m.r. spectrum.

Having induced crystallization at $-150°C$, Jensen separated the crystals from the solution. The solution now showed only the spectrum of one conformation. The crystals were redissolved at the same temperature of $-150°C$, giving a different n.m.r. spectrum characteristic of the second conformation. The two conformers of chlorocyclohexane can thus be isolated at $-150°C$. They have different physical properties, as is normal for isomers. The spectrum of each of these isomers changes as it is warmed, passing through the coalescence point and giving again the average spectrum of chlorocyclohexane at room temperature.

Freezing of conformational inversion in cyclohexane compounds

A reduction in temperature slows, then stops, the sequence of deformations undergone by cyclohexane, as shown in Figure 3.4 (p. 58). It would also be interesting to have available cyclohexanes that are rigid at ordinary temperature. The method proposed by Winstein, in 1955, consists of introducing the tertiary butyl group into the molecule. The tertiary butyl group is particularly large and cannot take up an axial position without considerably deforming the ring, which is energetically very unfavourable. The substituted cyclohexane, therefore, remains fixed in the conformation E:

E   A

Many physicochemical studies have been made on tertiary butyl cyclohexanes as simple models, using the hypothesis that the tertiary butyl group does not deform the ring and does not influence the phenomena observed. For example, the axial acetate (5) is hydrolysed with greater difficulty than the equatorial acetate (6), the difference in reactivity revealing the hindrance of the axial position.

(5)   (6)

The conformational preference of substituents

In mobile systems, either cyclic or acyclic, the conformation assumed by a molecule depends mainly on the repulsive or attractive nonbonded interaction between atoms or groups of atoms. Steric repulsion originated in van der Waals forces. The van der Waals radius of atoms or radicals is a parameter which allows steric hindrance to be estimated. The need was recognized for an experimental method for the comparison of the apparent steric effects of various groups in the cyclohexane series, as these steric effects are not always directly related to the van der Waals radii. The now classic work of the research groups of Winstein and Eliel has provided a solution to this problem. The preference for an equatorial conformation has been taken as a measure of effective size. In effect, a sterically demanding group in an axial position interacts with two hydrogen atoms, also axial, in positions 3 and 5, and the conformational equilibrium is displaced toward the equatorial conformation.

All the methods used amount to determining the conformational equilibrium constant K for the process A⇌E for monosubstituted cyclohexanes having a substituent R.

To determine K in the general case, a property (physical or chemical) of the mobile system is examined and compared with those found for two standards, one purely axial A′, the other exclusively equatorial E′.

A′ and E′ are cyclohexanes made rigid by a t-butyl group in position 4. This group is assumed to be sufficiently remote not to interfere in any way with R. One also assumes that the t-butyl group does not deform the ring, relative to the corresponding monosubstituted cyclohexane. These two hypotheses are reasonable, although the second has been questioned to some extent recently.

The use of n.m.r. is particularly convenient for studying the conformational equilibrium A⇌E. Consider the signal of a characteristic proton situated on carbon atom 1 or in the radical R of a monosubstituted cyclohexane. The chemical shift δ of the signal is intermediate between δ_a (conformation A) and δ_e (conformation E). If N_A and N_E are the molar fractions of these conformations, the chemical shift observed will be such that:

$$\delta = N_A\delta_a + N_E\delta_e.$$

Since $N_A + N_E = 1$ and $N_E/N_A = K$ this equation can be rewritten:

$$\delta = (\delta_a + K\delta_e)N_A \quad \text{with } N_A = 1/(1 + K),$$
$$\delta = (\delta_a + K\delta_e)/(1 + K), \text{ hence } K = \delta_a - \delta/(\delta - \delta_e).$$

δ_a and δ_e are measured using model compounds A′ and E′, and the equilibrium constant K can then be calculated.

The kinetic method presents some analogy with the n.m.r. method, in that it is also based on three experiments, the measurements of chemical shift being replaced by measurements of apparent reaction rate k for the cyclohexane derivative having one R substituent, present as a mixture of the conformations A and E. It is assumed that this rate constant is of the form (Winstein–Holness equation)

$$k = N_A k_a + N_E k_e,$$

Translator's footnote
The Winstein–Holness equation can, of course, be generalized by adding a term $N_F k_f$, to deal with the few cases in which for special reasons the flexible conformer makes an appreciable or even a predominant contribution to the total reactivity. The solvolysis of *trans*-3,5-di-*t*-butylcyclohexanol toluenesulphonate, investigated by Hanack, is a case where the reaction must take place mainly through a flexible conformer (perhaps rather distorted), since in either chair conformation one *t*-butyl group must be axial.

k_a and k_e being the rate constants appropriate for conformations A and E. These values, being unavailable because of interconversion of conformations A and E, are deduced from kinetic experiments with the rigid models A′ and E′. A calculation similar to that developed above leads to the relationship:

$$K = k_a - k/(k - k_e).$$

It should be emphasized that in the two methods which have just been discussed, the experiment is carried out on an averaged system. The calculation of the constant K of the conformational equilibrium is valid only when this equilibrium is more rapid than the method by which the mobile system is observed, as is certainly the case with n.m.r. spectroscopy or chemical reactivity[1].

A third method called the equilibrium method is of less use; it involves effecting the interconversion A′⇌E′ by a chemical route.

Table 3.1 *Differences in conformational free energy, ΔG^{0}, between equatorial and axial substituents (at 25°C except where otherwise shown)*

Substituent	$-\Delta G$ (kJ mol^{-1})	Solvent
CH_3	6.7–7.5	None
$CH(CH_3)_2$	7.5–10.0	—
CH_2CH_3	6.7–9.2	—
C_6H_5	11.0	Ether (35°C)
CO_2Et	5.0–5.8	Ethanol
OAc	1.5	None or 87% ethanol
OTs	2.9	87% ethanol
OH	3.3	75% acetic acid (40°C)
	4.0	Isopropanol (90°C)
	3.8	Water
	1.7	Carbon disulphide (20°C)
Br	2.9	None or 87% ethanol
Cl	2.1	—

In Table 3.1 are shown the conformational free-energy differences for some common substituents. There is, in general, good agreement between the results obtained by different methods. In the table the solvent in which the equilibrium has been studied is shown, since the solvent is involved indirectly in the equilibrium. For example, $-\Delta G^{\theta}$ for OH = 3.8 in water, as compared with 1.7 in carbon disulphide. In carbon disulphide the hydroxyl group is slightly solvated and seems 'smaller' than in water.

[1] Direct observation of conformers A and E is possible by infrared spectrography. The Franck–Condon principle applies, and each isomer has enough time to give its own spectrum before changing to the other.

Utilization of conformational free-energy differences

The conformational analysis of cyclohexane systems is greatly assisted by knowing the characteristic values of $-\Delta G^\theta$ for substituents. It can be accepted as a first approximation, that the values in Table 3.1 (applied to monosubstituted cyclohexanes) are additive in the conformational equilibria of polysubstituted cyclohexanes. The qualitative, and in many cases quantitative, prediction of a conformational equilibrium thus becomes possible. For example, cis-4-methylcyclohexanol, which exists in conformations (II) (CH_3 equatorial, OH axial) and (I) (CH_3 axial, OH equatorial):

(I)   (II)

In Table 3.1, $-\Delta G^\theta(CH_3)=7.5$, $\Delta G^\theta(OH)=3.3$, which shows that methyl has a preference for an equatorial conformation much stronger than that of the hydroxyl group. Therefore, an equilibrium displaced toward conformation (II) can be predicted. These values similarly allow an evaluation of the position of equilibrium. If K is the constant (at 25°C) for the equilibrium (I)$\rightleftharpoons$(II), that is to say:

$$K=[(II)]/[(I)],$$

then, by the hypothesis that free energies are additive, it can be calculated that:

$$\Delta G^\theta = -RT\ln K = -7.5-(-3.3)= -4.2\,\text{kJ mol}^{-1}.$$

The position of this equilibrium has been measured and is found to be $\Delta G^\theta = -5.0\,\text{kJ mol}^{-1}$. The agreement between the calculation and the experiment is not always as good as this when considering substituents 1,4 to each other. The effects of each substituent would be expected to be additive because they are relatively distant from one another; but this assumes a cyclohexane ring with geometry independent of the nature of the substitution. Such a hypothesis is not valid in every case, the cyclohexane ring being distorted slightly by the substituents which are attached to it. Naturally in these circumstances the exact addition of increments of $-\Delta G^\theta$ is not possible. One can easily imagine that an axial group may cause distortions of the ring which hinder an axial substituent at position 4. The calculations of the conformational equilibria are thus in error, because the increments used which are characteristic of monosubstituted cyclohexanes do not take account of the reciprocal effect. The distortion of the cyclohexane ring, or rather its susceptibility to distortion, similarly explains the exceptions to the addition of the $-\Delta G^{\theta}$ values for 1,2 di-substituted cyclohexanes. The dipolar interactions between substituents should also be considered in predicting a conformational equilibrium.

Cyclobutane

For a long time cyclobutane was presumed to be planar. It is now known from crystallographic and spectroscopic studies that this is true only for certain particular cases; usually cyclobutane is more or less folded. This distortion of the ring gives an axial and equatorial character to the substituents. It is thus easy to explain that the *cis* disubstituted 1,3 cyclobutanes are more stable than their *trans* isomers:

trans   *cis*

With a planar ring one would expect greater stability for the *trans* stereoisomer, in which 1,3 interactions between the groups are absent. The folded conformation avoids a direct interaction between 1,3 substituents, and optimal stability is achieved when the two substituents are pseudo-diequatorial, which is possible only in the *cis* isomer.

Cyclopentane

The planar conformation is attributed to cyclopentane if one treats the angular strain as the only relevant factor. In the regular pentagon each interior angle is 108°, very close to the value of 109° for tetrahedral carbon. Indeed, construction of a model of cyclopenane with sp^3 carbon atoms of the Dreiding type leads to an almost planar ring. It is now known that the real situation is much more complex. The planar conformation of cyclopentane is disfavoured by torsion forces resulting from the multiply eclipsed bonds. Cyclopentane changes rapidly between nonplanar conformations of which the most characteristic are the envelope conformation (having four coplanar atoms) and the half-chair (having only three coplanar atoms):

Envelope   Half-chair C_2

There is no marked energy minimum for the family of nonplanar conformations of cyclopentane, which thus shows great flexibility. The term pseudorotation is used to describe the incessant fluctuation of the fold of the envelope or the C_2 axis of the half-chair. The conformational analysis of cyclopentane derivatives is a difficult problem; according to the nature of the substituents it is necessary to consider either an envelope or a half-chair conformation.

Rings larger than cyclohexane

Cycloheptane is a relatively flexible system in which there are two groups of flexible conformation, chair and boat, separated by an interconversion barrier of the order of $34 \, kJ \, mol^{-1}$. The ideal chair form in Figure 3.6

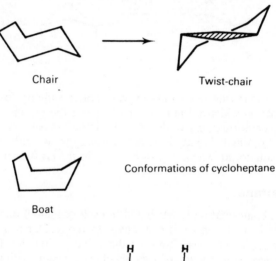

Chair   Twist-chair

Conformations of cycloheptane

Boat

Cyclodecane

Figure 3.6

is capable of deformation without much variation in potential energy. The most stable chair form is that in which torsion interaction is minimized by means of a rotation about the C_4—C_5 bond, leading to a twist-chair conformation. The cycloheptane boat similarly deforms by pseudorotation, giving a group of boat conformations.

The privileged conformations of large rings are still not well known; for example, those of cyclodecane, $C_{10}H_{20}$, for which the crystalline structure has been determined by X-ray crystallography through the work of Dunitz.

The general shape of the molecule is shown in Figure 3.6, where one can see units of *anti* butane and also units of *gauche* butane. Some of the hydrogen atoms are compressed in the interior of the ring, leading to slight deformations.

Introduction of an sp² carbon atom into a ring

Cyclohexane only will be discussed here. The general shape of the ring is not appreciably altered by the introduction of one sp^2 carbon atom, and remains in the chair conformation:

Cyclohexanone   Methylenecyclohexane

There is, however, some difference between cyclohexane and cyclohexanone; for example, the trigonal carbon atom has led to the disappearance of an axial linkage. An axial substituent, placed 1,3 to the ketone function, therefore has one interaction with an axial hydrogen atom, compared with two in cyclohexane. The equatorial preference for a substituent is thus diminished by the presence of a ketone function in the position 1,3 to it (3-alkyl ketone effect):

An exocyclic carbon–carbon double bond hinders an equatorial substituent on the adjacent carbon atom. This *allylic strain*, or $A^{1,3}$ strain, according to the terminology proposed by Johnson and Malhotra, results from the eclipsing interaction between the double bond and the equatorial substituent. The $A^{1,3}$ strain can be strong enough to displace the conformational equilibrium toward the axial conformation:

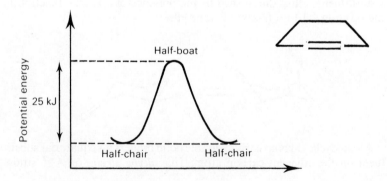

Cyclohexene

The introduction of a double bond into cyclohexane alters the geometry of the ring. The most stable conformation is called the half-chair. One can distinguish axial linkages (*a*) and equatorial linkages (*e*) for the homoallylic carbon atoms. The axial and equatorial character is less marked for the allylic positions, which are designated pseudo-axial (*a'*) and pseudo-equatorial (*e'*).

Half-chair   Half-chair

Cyclohexene undergoes an easy inversion of the ring, the inversion barrier consisting of the half-boat conformation:

Potential energy

25 kJ

Half-boat

Half-chair   Half-chair

The cyclohexene structure is found in many organic molecules. The presence of a double bond causes characteristic steric effects, some of which will be discussed later. However, it must be noted that a pseudo-equatorial substituent R is hindered by a group R′ situated on the double bond. If R and R′ are sufficiently large the allylic strain which is caused ($A^{1,2}$ strain) may render the conformation in which R is pseudo-axial more stable.

The bond to R(a′) forms an angle much larger than that of R(e′) with the plane of the double bond; the interference of the groups R and R′ is thus much reduced.

Six-membered heterocycles

The geometry of the cyclohexane ring is retained when one of the carbon atoms is replaced by a heteroatom such as oxygen, sulphur or nitrogen.

Tetrahydropyran   Piperidine

Oxygen heterocycles play a particularly important role in biochemistry, as they constitute the skeleton of many sugar molecules. It is interesting to note that Haworth introduced the term 'conformation', in 1929, in connection with the description of the structure of sugars.

The equatorial preference of an oxygen substituent often disappears when it is attached at the position next to the heterocyclic oxygen atom. The *anomeric effect* which is the origin of this anomaly is undoubtedly

Anomeric effect

due to unfavourable dipolar interactions[1]. In the diagram shown at foot of p. 69, assuming that the ring is prevented from inverting by suitable substitution, the anomeric effect will make the axial epimer the more stable.

Some fused polycyclic systems

The fusion of rings leads to innumerable conformations of varied shape. Structures built up from six-membered rings will be mainly dealt with here. Decalin consists of two hexagons fused at one side (Figure 3.7). Two stereoisomers are known, *trans*-decalin (7) and *cis*-decalin (8), *cis* or *trans* describing the relative positions of the two substituents at the ring junction ('angular' substituents).

The most stable form is that in which each ring can adopt a chair conformation. In *trans*-decalin (7) the angular hydrogen atoms are axial relative to both the rings. In *cis*-decalin (8) each of the two hydrogen atoms is simultaneously axial for one ring and equatorial for the other. It is also useful to consider the axial or equatorial nature of the first atoms of each ring. For example, if one considers that *trans*-decalin (7) is a 1,2-disubstituted cyclohexane A, the substituents are joined in order to make ring B. Each substituent which begins the ring is equatorial. Such is not the case with *cis*-decalin (8); there one bond is axial whereas the other is equatorial. In this situation *trans*-decalin will be more stable than *cis*-decalin, having the larger number of equatorial substituents.

Finally, it should be noted that *trans*-decalin is a completely rigid system, the inversion of the rings being impossible (this can be demonstrated easily by making a molecular model). On the contrary, *cis*-decalin transforms into its enantiomer, by inversion of both rings. The n.m.r. spectrum at room temperature shows clearly the flexibility of *cis*-decalin, having equivalent axial and equatorial protons, and the rigidity of *trans*-decalin which gives distinct signals for axial and equatorial protons.

The perhydrophenanthrene system is found in many natural products. The *trans, anti, trans*-isomer (9) and the *cis, anti, trans*-isomer (10) are shown in Figure 3.7. Generally chair conformations are adopted, but in particular cases a boat conformation is necessary to relieve angular or steric strain. For example, it is easy to show that the *trans, syn, trans*-perhydrophenanthrene (11) cannot have all its rings in the chair form. If one attributes to the rings A and B the conformation of *trans*-decalin (7), it can be seen that the construction of ring C requires the use of two adjacent bonds, axial with respect to ring B, which is geometrically impossible. The molecule (11), therefore, exists with a non-chair conformation for ring B.

[1] For a recent discussion, and a new interpretation, see S. David, O. Eisenstein, L. Salem and R. Hoffmann, *J. Amer. Chem. Soc.*, 1973, **95**, 3806.

(7)

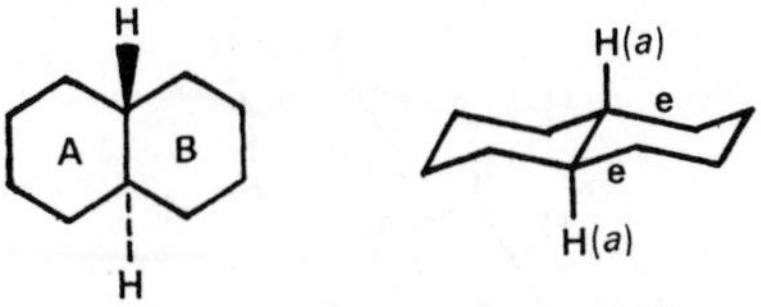

(8)

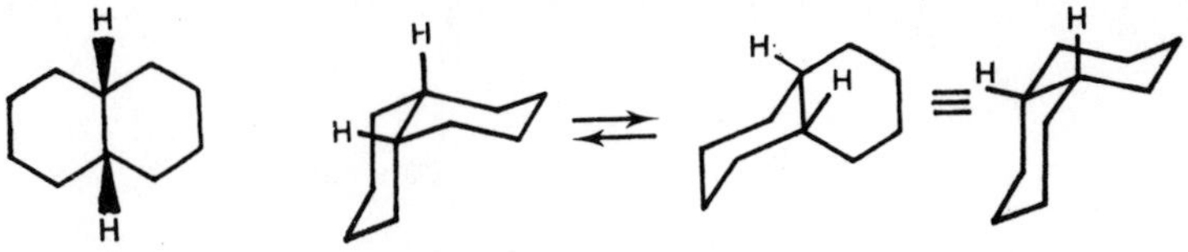

Decalins

(9)

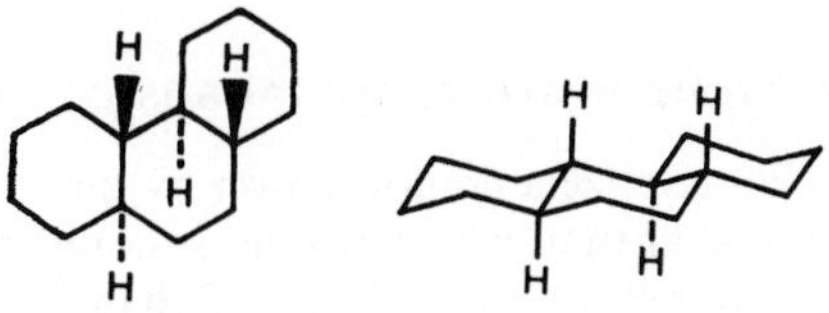

(10)

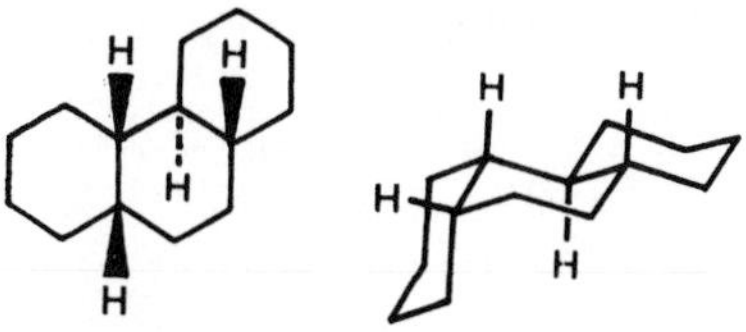

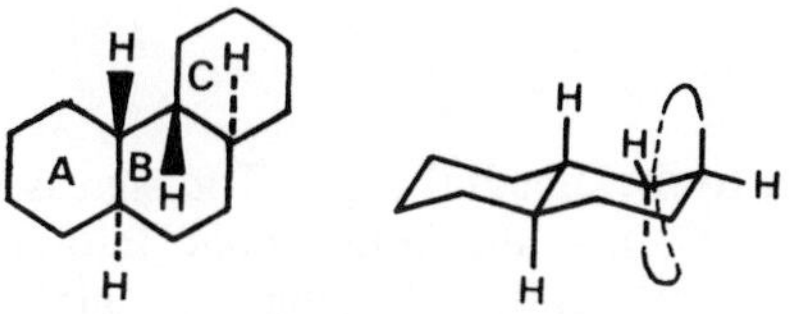

Perhydrophenanthrenes

Figure 3.7

This list of polycyclic molecules comprised of cyclohexane rings will not be continued, but bridged molecules (*see* Chapter 1, p. 28) will be discussed. The bridge can be a source of strain, imposing a particular geometry on the molecule. For example, in camphor, which is a rigid molecule, one can recognize a cyclohexane ring frozen into a boat form and two cyclopentane rings similarly frozen in an envelope conformation. In Chapter 2 (p. 42) the lactonization of a hydroxy acid, which forces the molecule to take up a boat conformation, was discussed.

Camphor

Adamantane

The existence of a bridge need not imply a non-chair conformation for six-membered rings. Adamantane is a particularly stable molecule in which all the cyclohexane rings are chairs.

Analysis of cyclic systems by means of dihedral angles

It is usual to characterize a conformation by the dihedral angles along the various 'pivots' around which rotations could occur to change the conformation. In this way the C—C—C—C chain of butane in the *gauche* form (12) defines a dihedral angle of 60°:

(12)

A sign is given to the dihedral angle. The sign is positive if rotation in a clockwise sense brings the frontal bond into coincidence with the rear bond. With this convention, the dihedral angle of *gauche* butane (12) is negative; it is easy to check that this sign is independent of the end of the molecule from which one observes it. The dihedral angle for rings is similarly defined. In chair cyclohexane, observed by a Newman projection (p. 45), two dihedral angles interior to the ring can be seen. They measure approximately 60° and are of opposite signs. The determination of the dihedral angles of rings identifies their conformation and their symmetry elements. An analysis of conformations by means of the dihedral angle has been published by Bucourt[1], and facilitates rapid qualitative discussion of many problems which arise in polycyclic systems. The principle of the method lies in the knowledge of the dihedral angles of some

[1] R. Bucourt, *Bul. Soc. Chim.*, 1964, 2080.

characteristic conformations of single rings (*see* Figure 3.8), and in the estimation (by calculation) of the deformation energies of the dihedral angles of these conformations. It has, for example, been calculated that a deformation of $\pm 10°$ in a dihedral angle of chair cyclohexane increases the potential energy of the system by $4\,kJ\,mol^{-1}$. The chair is thus capable of deformation without much difficulty. If an axial substituent R is situated at the hinge of a dihedral angle which is deformed, the decrease in the dihedral angle (closing it) is easy, because it reduces the interaction energy of the substituent with the axial 1,3 hydrogen atoms. However, the opening of the dihedral angle θ is more difficult because of the increased 1,3 diaxial strain which results:

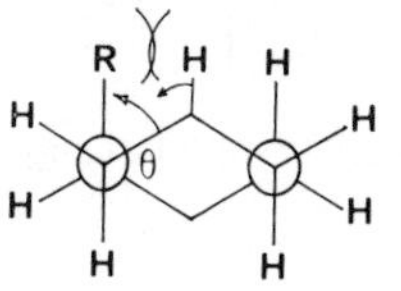

The deformation of a dihedral angle of cyclohexene in the half-chair form requires an increase in energy of the same order as that of the chair.

In bicyclic systems it is useful to consider the two dihedral angles at the junction of the rings (junction dihedrals). Figure 3.9 shows, in the Newman projection, a *trans* and a *cis* junction between the two rings. The dihedral angles of the *trans* junction are of opposite signs and the dihedral angles of the *cis* junction are of the same sign. If one deforms the dihedral θ_A by a rotation of the front carbon atom (for example, increasing θ_A in absolute value) the neighbouring dihedral angle θ_B is deformed: a reduction of θ_B if the junction is *trans*; an increase in θ_B if the junction is *cis*. The *trans* junction transmits a deformation of dihedral angle in an inverse direction, a *cis* junction transmits a deformation of dihedral angle in the same direction (in absolute value).

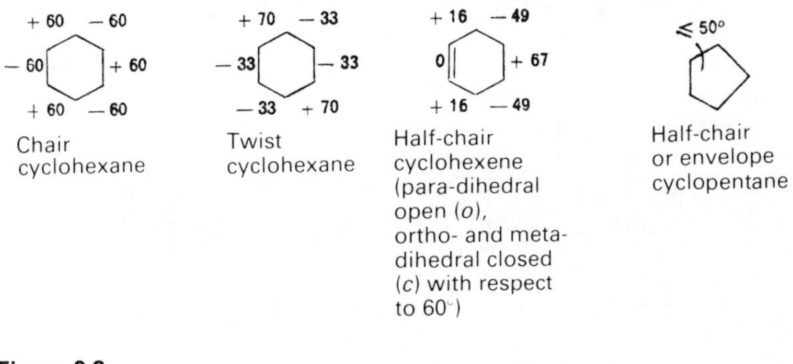

Chair
cyclohexane

Twist
cyclohexane

Half-chair
cyclohexene
(para-dihedral
open (*o*),
ortho- and meta-
dihedral closed
(*c*) with respect
to 60°)

Half-chair
or envelope
cyclopentane

Figure 3.8

$|\theta_A|\nearrow$ $|\theta_B|\searrow$

or

$|\theta_A|\searrow$ $|\theta_B|\nearrow$

The *trans* junction transmits a change in dihedral angle in an inverse direction

$|\theta_A|\nearrow$ $|\theta_B|\nearrow$

or

$|\theta_A|\searrow$ $||\theta_B|\searrow$

The *cis* junction transmits a change in dihedral angle in the same direction

Figure 3.9

The behaviour of many polycyclic systems can be rationalized using these ideas. We shall consider some examples. It is known that the diene (13) is more stable than its isomer (14), the equilibrium between the two compounds being displaced toward (13). To understand this phenomenon, it may be assumed arbitrarily that the bicyclic system is obtained by the fusion of two half-chair cyclohexene rings. Therefore, in which of these isomers does this fusion require the minimum deformation of the rings?

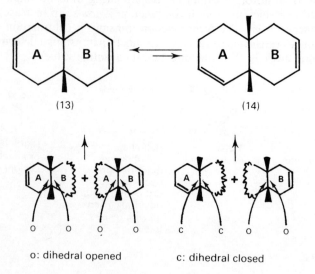

(13)   (14)

o: dihedral opened   c: dihedral closed

The fusion of two cyclohexene rings to give (13) can be carried out without difficulty. Thus, in ring A, the dihedral opposite the double bond is 'open', which requires a similar opening of the dihedral of ring B at the opposite side of the junction (a *cis* junction). This last dihedral is situated opposite the double bond of ring B, which is the best possible position in which to have an opened dihedral. There is, therefore, no additional angular strain.

Now consider the geometry of (14). The dihedral at the ring junction in ring A is 1,3 to the double bond, that is a closed dihedral. The *cis* junction requires a closed junction dihedral in ring B, which is incompatible with the position of the double bond in ring B. The rings A and B in (14) thus cannot have the geometry of half-chair cyclohexene, so some deformation is unavoidable and renders 14 less stable than its isomer (13).

A similar form of reasoning explains why the *trans*-tetrahydroindanone (15) chlorinates preferentially at the indicated position:

(15)

The chlorination of a ketone involves an enol as intermediate. The direction of enolization will be determined by the need to minimize strain in the enol. It has been seen that the dihedral angles of cyclopentane in a half-chair or envelope conformation can be regarded as closed (Figure 3.8). To comply with this requirement, the double bond of the enol chooses the position opposite the ring junction, thus the dihedral of the junction in the cyclohexene ring opens, and the *trans* junction transmits a closure to the cyclopentane ring. A change in the site of halogenation is predicted for a *cis* hydrindanone, and is found by experiment.

There are many examples where analysis by dihedral angle can provide useful information on the behaviour of polycyclic systems. Deformations induced in the steroid skeleton by the introduction of a double bond are in most cases predictable by the analysis of dihedral angles.

Conclusion

Conformational analysis is an area of stereochemistry which has attracted much attention from chemists in the last twenty years and has been honoured by the Nobel Prize in Chemistry for 1969 (Barton and Hassel).

There is a great potential for conformational analysis in cyclic compounds, and a need for consideration of all the structural characteristics (double bonds, ring fusions, etc.) in order to make valid comparisons between the behaviour of different molecules. Although it is common to find cyclic structures rigid or conformationally well defined, this is not the case with acyclic compounds, which are almost always mobile and conformationally heterogeneous. The study of conformational equilibrium in solutions of acyclic compounds is generally carried out indirectly by means of nuclear magnetic resonance. Thus the coupling constant J

of vicinal protons, $\begin{array}{c} H \quad H \\ | \quad | \\ C{-}C \end{array}$, depends on the angle θ of the dihedral

$H{-}C{-}C{-}H$. Knowing the relationship $J = f(\theta)$ (Karplus relationship) and the experimental J value, which represents a mean, it is possible to deduce the forms which predominate in a conformational equilibrium. X-ray crystallography gives a true photograph of the molecule in the shape that it adopts in the crystalline lattice. It is useful to assume that cyclic compounds of little mobility do not change in conformation on passing from the crystalline to the dissolved state, for example, cyclodecane (p. 66). Such a hypothesis is debatable for flexible molecules. In general, compounds crystallize in a single conformation, which is not necessarily the most stable in solution. However, X-ray crystallography has been exceedingly useful to conformational analysis, in showing the stereochemist the conformations present in the solid, and therefore allowing him to argue by analogy for the liquid or dissolved state. It should be noted that the difficult problem of the active sites of enzymes is at present being tackled by X-ray crystallography on crystalline enzymes. The tertiary structure of proteins is thus defined. For example, Kendrew has determined the structure of myoglobin, a protein having a polypeptide chain of 153 amino acids.

Electron diffraction measurements on gaseous compounds can provide information about privileged conformations in the gas phase. The method has, as yet, been relatively little used.

Some of the basic principles of conformational analysis have been considered, confined to the simple case where the conformation is essentially the result of a compromise between torsion energy, Baeyer strain and steric interactions. Naturally, additional effects may come into play. There is conjugation, a parameter which tends to favour coplanarity. Dipolar interactions (*see* the anomeric effect, p. 69) frequently intervene in an important manner in a conformational equilibrium. For example, 2-bromocyclohexanone exists largely with the halogen atom axial, to avoid a dipole–dipole interaction:

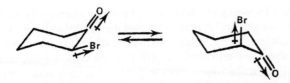

Hydrogen bonds can give a particular conformation to a molecule in which, *a priori*, internal rotations could exist. A good example is the helical conformation imposed on polypeptide chains (Figure 3.10) by the repetition of hydrogen bonds between the functions NH and C=O, four peptide linkages apart. The amide linkages are planar in the macromolecule.

Figure 3.10
α-Helix of polypeptides (hydrogen bonds ... are not all shown)

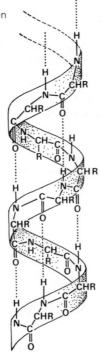

The complexity of the problems introduced by conformational analysis has required the use of methods of calculation involving computers. They have proved very useful in the calculation of the energy of many conformations, so that the preferred conformation(s) can be selected. These methods were originally purely empirical, treating the molecule as a mechanical system subject to constraints of torsion and strain of which the laws are known. Calculations are made by altering the geometrical parameters (angles and distances) in seeking energy minima. Along with these calculations by molecular mechanics, which are proving to be very accurate, there are now semi-empirical or *ab initio* calculations, using quantum mechanics, which seem to promise a much wider application.

Finally, it must be emphasized that the principles of conformational analysis are indispensable not only for understanding the shape of a molecule but also for predicting its chemical reactivity. Conformational effects appear in the transition states of reactions, a factor which the chemist must take into account and which will be discussed in Chapter 5, on dynamic stereochemistry.

Bibliography

For an introduction to conformational analysis, see:

E. L. Eliel, *Elements of Stereochemistry*, Wiley, 1969.
K. Mislow, *Introduction to Stereochemistry*, W. A. Benjamin, 1965.
G. Natta and M. Farina, *Stereochemistry*, Longman, 1972.

For a specialized work, see:

E. L. Eliel, N. L. Allinger, S. J. Angyal and G. A. Morrison, *Conformational Analysis*, J. Wiley and Sons, 1965.

4

Stereoisomerism

Introduction

The idea of isomerism is of primary importance in organic chemistry. It owes its origin to the fact that a given molecular formula can correspond to several compounds (isomers), which possess different properties. The molecular formula states the nature and the number of atoms in the molecule; this information is insufficient to define completely a chemical compound.

The term 'structural isomers' is used for compounds having the same molecular formula but differing in the sequence of atoms or linkages between the atoms. For example, there are two olefins of the molecular formula C_4H_8, but-1-ene, $CH_3-CH_2-CH=CH_2$, and but-2-ene, $CH_3-CH=CH-CH_3$. The structural formulae given, which indicate the terminal or internal character of the double bond, do not completely describe all the possibilities of isomerism. In fact, two isomeric but-2-enes are known. The existence of this type of isomerism can only be understood by examining the position of the atoms in space, using molecular models; for this reason isomers of the type are called stereoisomers. In the present example, the stereoisomerism involves the planarity of the double bond, which makes possible two relative arrangements for the methyl groups:

cis-but-2-ene   *trans*-but-2-ene

The non-equivalence of *cis* and *trans*-but-2-ene is easily confirmed by the impossibility of superimposing these two molecules on top of one another, and by their different reactivity.

Another example of stereoisomerism is the molecular formula C_5H_{10}

to which there correspond two isomeric dimethylcyclopropanes, (1) and (2):

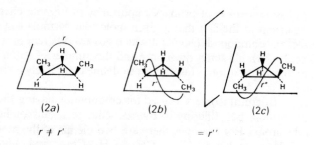

(1)   (2)

1,2-dimethylcyclopropane (2) consists of a ring, necessarily planar, and exists in the form of three stereoisomers: (2a), (2b), (2c).

(2a)   (2b)   (2c)

$r \neq r'$ $= r''$

If one carefully examines these molecules, which are stable and isolable compounds, one can see that two types of stereoisomers are involved. Comparing (2b) with (2c) it can be seen that all the distances between non-bonded atoms in one stereoisomer are similar to those in the second stereoisomer (for example, the distance r' or r'' between the two carbon atoms of the methyl groups). The cyclopropanes (2b) and (2c) are isometric. However, they are different, and it is impossible to superimpose them. By appropriately arranging the two molecules it can be seen that they are images of one another reflected in a mirror. Therefore, (2b) and (2c) are two mirror images and not superimposable.

However, the difference between (2a) and (2b) or (2c) is quite clear, because the two compounds to be compared are not isometric, and there is here no question of structural identity. For example, in (2a) the two methyl groups are relatively close together (cis stereochemistry) whereas in (2b) (trans stereochemistry) the two methyl groups are further apart. That is to say $r < r' = r''$.

Classification of stereoisomerism

Mislow's classification of stereoisomerism, adopted here, is based on the fundamental properties which have just been exemplified in the case of the dimethylcyclopropanes. Two stereoisomers which are mirror images will be called enantiomers, two stereoisomers which are not enantiomers will be called diastereoisomers (or diastereomers). When comparing two molecules A and B deriving from a single constitutional isomer, an operation which is fundamental to the idea of isomerism, it is important to see immediately whether one is converted to the other by reflection in a mirror. If so, one must first check that the object A and its image B are not identical before calling them enantiomers. For example, (2a) and (3a) are object and mirror image with respect to a plane parallel to the cyclopropane ring, but these molecules are in fact superimposable after the rotation of (3a), and therefore identical.

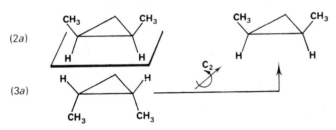

The steric relationships between molecules can be summarized using the following table:

B is the mirror image of A $\Rightarrow$ $\begin{cases} \text{B is identical to } A, \text{ or} \\ \text{B is the enantiomer of A} \end{cases}$

B is not the mirror image of A $\Rightarrow$ B is the diastereoisomer of A

It follows, therefore, that two enantiomers are not at the same time diastereomers of one another, and two diastereomers cannot be enantiomers of one another. However, it is possible that two diastereomers each possess, in addition, an enantiomer. One enantiomer can similarly be related

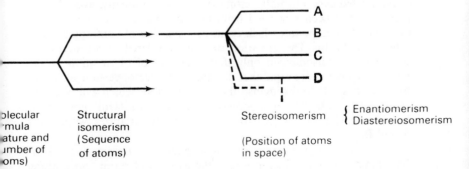

Figure 4.1   Relationships between isomers

Example:

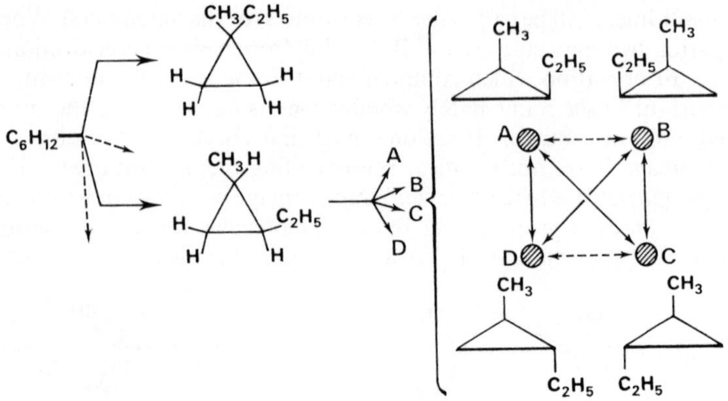

A is enatiomeric (◀---▶) with B, and diastereoisomeri (◀——▶) with C or D

Figure 4.1   Relationships between isomers—contd.

to several diastereomers. These relationships are represented in Figure 4.1. The example used is the general one of 1-methyl-2-ethylcyclopropane, which leads to four stereoisomers.

General definition of isomerism

It has been assumed so far that isomers, which by definition have the same molecular formula, are individual chemical species which can be separated one from another. This last criterion is useless, although it is always retained by chemists who are synthesizing stable and isolable compounds. Compounds will now be considered as isomers if they have been characterized experimentally, either directly or indirectly, even though they have not been separated from one another. Let us consider, for example, the compound which exists as a mixture of conformations in rapid equilibrium. This situation can be considered as an additional source of stereoisomerism only if the method of observation of the mixture is much more rapid than the speed of interconversion, as it thus becomes possible to distinguish the conformations which will be called stereoisomers. The notion of isomerism only assumes its full meaning for the chemist when an energy barrier, however low, separates two isomers A and B.

When the energy barrier for isomerization is greater than about $100 \, kJ \, mol^{-1}$ it is possible to isolate the isomers at room temperature. To isolate or demonstrate the existence of isomers which are separated

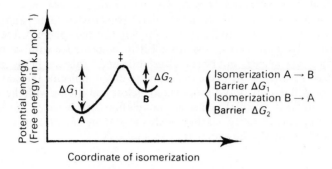

Coordinate of isomerization

{ Isomerization A → B
  Barrier ΔG_1
  Isomerization B → A
  Barrier ΔG_2

by a barrier lower than that, it is necessary to reduce the temperature in order to slow down the isomerization. One can also use spectroscopic methods especially adapted either to the study of the rate of isomerization or to the observation of the isomers themselves.

Enantiomerism

The relationship between two enantiomers is that between an object not superimposable on its mirror image, and that image. The 'chirality' (from the Greek *cheir*, hand) is the geometrical property which characterizes non-identity of an object and its mirror image.

A sphere is superimposable on the image of a sphere reflected in any plane. A translation is enough to effect coincidence. On the other hand, the same operation is impossible with a glove or a corkscrew. The image of a right-hand glove is a left-hand glove, and it is not difficult to convince oneself of the non-identity of these two gloves, if only because one cannot place the right hand in the left glove.

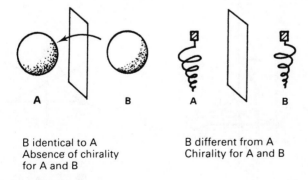

B identical to A
Absence of chirality
for A and B

B different from A
Chirality for A and B

Any chiral substance has at least one stereoisomer, its enantiomer. For example, *trans*-1,2-dimethylcyclopropane (2*b*) is chiral and (2*c*) is an enantiomer (or antipode). These two isomeric molecules play roles similar to those of two corkscrews A and B. *cis*-1,2-Dimethylcyclopropane (*2a*) is identical to its mirror image (*3a*) (*see* p. 80), like the sphere A, the only difference lying in the fact that to effect its coincidence with its image it is necessary to carry out both a translation and a rotation (if the mirror is placed parallel to the plane of symmetry of the molecule the superimposition will require only a translation). Before continuing with the study of chirality it is essential to discuss the relationships between molecular symmetry and chirality, starting with the system of notation used by chemists to specify the symmetry of a molecule.

Point groups and symmetry

It is obvious when one solid is more symmetrical than another, but it is essential to define two concepts precisely; those of symmetry operations and of elements of symmetry. The symmetry operation is the movement of an object which brings it back into coincidence with itself when the movement is finished. There is then no way of knowing if the object has or has not been submitted to a symmetry operation. The elements of symmetry of a solid (or of the molecule in the present example) are the geometric elements in relation to which the symmetry operations are carried out. These elements can be a point, an axis or a plane.

The plane of symmetry, σ, or mirror plane, divides a solid into two parts, one the image of the other. The corresponding symmetry operation consists of taking all the points of the solid and moving them to symmetrical positions, with respect to the plane. The planes passing through the poles of the sphere are symmetry planes for it. Cyclopropane possesses three vertical planes of symmetry (σ_v) and one horizontal plane of symmetry (σ_h) (Figure 4.2):

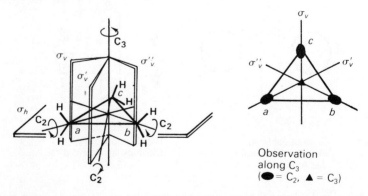

Observation along C_3
($\bullet$ = C_2, $\blacktriangle$ = C_3)

Figure 4.2   Symmetry elements of cyclopropane

Trans-1,2-dimethylcyclopropanes (2*b*) and (2*c*) have no plane of symmetry, whereas *cis*-1,2-dimethylcyclopropane (2*a*) has retained one of the symmetry planes of cyclopropane.

An axis of symmetry, C_n, defines a rotation of the object around this axis. It is usual to orientate a molecule so that the axis of highest order is placed vertically. Returning to the example of cyclopropane, it can be seen that a rotation of $2\pi/3$ around the C_3 axis perpendicular to the plane of the ring has the effect of permuting each corner with its neighbour; *a*, *b* and *c* move to occupy the positions *b*, *c*, *a*, if the direction of rotation is clockwise. At the end of the operation the molecule occupies a position indistinguishable from its original position. The rotation around a C_2 axis is shown by the interchange of two of the three atoms of cyclopropane. For example, rotation around the C_2 axis going through the atom *a* leaves that atom unchanged, whereas the atoms *b* and *c* are interchanged.

An improper axis of symmetry results from the combination of two symmetry operations, a rotation of order *n* followed by a reflection in a plane perpendicular to the axis of the rotation. If these two operations, carried out successively, bring the object back to its starting configuration it is described as being characterized by an axis S_n. There are three S_4 axes in methane, coinciding with the C_2 axes. It is important to note that the two operations which allow the definition of S_n involve elements (axis and plane) which are not necessarily elements of symmetry of the molecule. The plane $\sigma_\perp$ involved in the operation S_4 is not a plane of symmetry

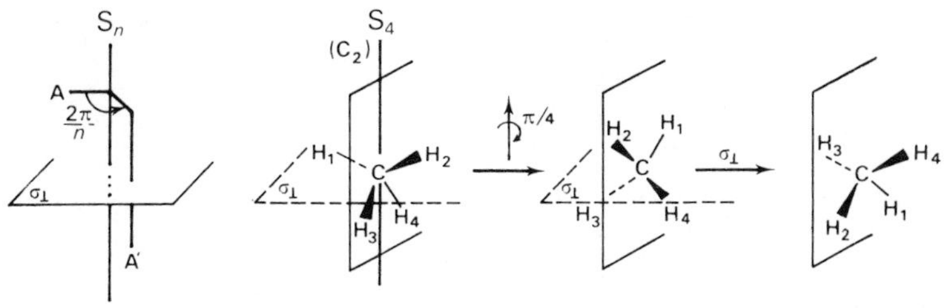

Asis S_n: A → A'   Axis S_4 in CH_4

Figure 4.3

of methane. It is nevertheless evident that a molecule having both an axis of symmetry C_n and a plane of symmetry perpendicular to this axis necessarily possesses an axis of improper symmetry S_n. In cyclopropane (Figure 4.2) the C_3 axis is perpendicular to the plane of the ring σ_h, which defines a vertical improper axis S_3.

Centre of symmetry   Some molecules have a structure such that it is possible to find for each atom an equivalent atom by joining the first atom to
a point (*i*) and prolonging the line by an equal length. This point is
called a centre of symmetry of the molecule. In the tetramethylcyclohexane

Table 4.1   *Some symmetry point groups*

Symbol of group	Principal elements of symmetry	Examples
C_1		$CHO - CHOH - CH_2OH$ $\overset{X}{\underset{H}{>}} C = C = C \overset{Y}{\underset{H}{<}}$
C_s	σ ▱	$\overset{Cl}{\underset{Cl}{>}} C = C \overset{H}{\underset{Cl}{<}}$
$S_2 \equiv C_i$ S_n	S_2 (or i) $S_n, C_{n/2}$ ↑	 $Cl \cdot \overset{Me}{\underset{Me}{}} \cdot Cl$ (with H's)
C_n	C_n ↑ C_n	C_2 $\overset{X}{\underset{H}{>}} C = C = C \overset{X}{\underset{H}{<}}$
C_{nv}	C_n σ_v ∧ σ_v $n\sigma_v$	C_{3v} $\overset{}{Cl} \overset{H}{\underset{Cl}{\wedge}} Cl$ C_{2v} $CH_3 \quad CH_3$
C_{nh}	C_n σ_h ▱ σ_h	C_{2h} , $\overset{}{\underset{CH_3}{\Big/}} \!\!\!=\!\!\!\bullet\!\!\!=\!\!\! \overset{CH_3}{\Big\backslash}$

Table 4.1   *Some symmetry point groups—contd.*

Symbol of group	Principal elements of symmetry	Examples
D_n	C_n $nC_2 \perp C_n$	
D_{nd}	C_n $nC_2 \perp C_n$ $n\sigma_v$	
D_{nh}	C_n $nC_2 \perp C_n$ $n\sigma_v$ σ_n	
T_d	$4C_3$ $3C_2$ 6σ	

(*4*) the centre of symmetry is the only element of symmetry, except for an S_2 axis which is equivalent to it[1].

(4)

Note that the pairs of hydrogen atoms and the methyl groups joined to the atoms of the ring and symmetrical with respect to the centre (*i*) are placed at positions 1,4 one to another, some of the groups being on one side and some on the other of the average plane of the cyclohexane ring.

Symmetry groups

When considering the symmetry elements of cyclopropane (Figure 4.2) it was seen that the presence of some symmetry elements necessarily implies the presence of other symmetry elements. For example, an S_3 axis is the direct consequence of the simultaneous presence of a C_3 axis and a plane σ_h ($S_3 = C_3 \times \sigma_h$). The intersection of planes of symmetry σ_h and σ_v creates a C_2 axis ($C_2 = \sigma_h \times \sigma_v$)[2].

It is convenient to describe a molecular structure by naming the minimum elements of symmetry capable of generating the set of symmetry elements. Mathematicians, in developing group theory, have been able to deduce what combinations of symmetry operations are possible, and what symmetry operations are a consequence of other symmetry operations. For solids, of which the forms are the most varied, theoretically an infinite number of point groups can exist, that is to say of combinations in which the symmetry elements all go through the same point (which is not necessarily within the solid). Only the case of the isolated molecule, as distinct from that of the crystal lattice, will be considered here.

Table 4.1 shows some characteristic groups, in particular those which are useful for recognizing molecular chirality. The symbols of these

---

[1] An axis S_1 is equivalent to a plane of symmetry σ. It is always possible to define a symmetry operation which is the product of a rotation of order 1 around an axis perpendicular to the plane σ (a complete rotation leaving the molecule unmoved) followed by a reflection with respect to the plane.

[2] Referring to Figure 4.3, where a diagram defines the axis S_n, it can be seen that for $n = 2$, A and A′ can similarly be related one to another by a reflection through the point of intersection of the axis and the plane.

groups show, as far as possible, the symmetry elements of the group. C_{nv} is associated with the presence of an axis C_n and a plane σ_v, C_{nh} is the group which contains an axis C_n and a plane σ_h. In theory, the number n may have any value but in practice it remains small. Nevertheless, $n = \infty$ is found sometimes in very symmetrical molecules.

Assigning a symmetry point group to a molecule is relatively easy if one first looks for the symmetry axes.

(a) A symmetry axis of order infinity is characteristic of linear molecules ($D_{\infty h}$) and ($D_{\infty v}$). Acetylene, $D_{\infty h}$, is represented below with some of its symmetry elements (the planes σ_v are not indicated).

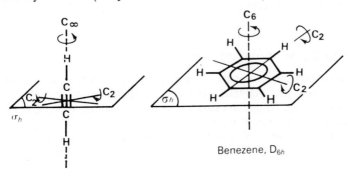

Benezene, D_{6h}

Acetylene, $D_{\infty h}$

(b) Having verified the absence of a C_∞ axis, one checks to see if there are several axes of order greater than 2. If so, one is dealing with a molecule which is still very symmetrical, such as CH_4, with the symmetry of the tetrahedron T_d ($4C_3$, $3C_2$, 6σ) (Figure 4.3). Three S_4 axes are also present.

(c) In the case of an axis of symmetry C_n ($n \geqslant 2$) associated with n C_2 axes perpendicular to C_n, the symmetry group can only be D_n, D_{nh} (a plane σ_h) or D_{nd} (n planes σ_v). Benzene possesses a C_6 axis perpendicular to the ring, which contains six C_2 axes. The ring being of planar symmetry σ_h, benzene belongs to the group D_{6h}.

(d) If the compound examined contains only a single axis C_n as its symmetry element, it is of the group C_n. For example, the allene

$$\begin{array}{cc} H & H \\ & \diagdown \quad \diagup \\ & C{=}C{=}C \\ & \diagup \quad \diagdown \\ Cl & Cl \end{array}\quad \text{has symmetry } C_2.$$ The additional presence of an im-

proper axis S_n, a plane σ_v or σ_h will allocate it respectively to the symmetry groups S_n, C_{nv} or C_{nh}. The group S_2 is equivalent to C_i (a centre of symmetry). The tetramethylcyclohexane (p. 88) is an example of a structure S_2 or C_i.

A C_n axis and n planes σ_v are characteristic of the group C_{nv}. For example, methyl chloride, CH_3Cl, belongs to the group C_{3v}, the C_3 axis passing along the CCl linkage. *cis*-But-2-ene (p. 79) belongs to the C_{2v} group, although *trans*-but-2-ene is of the group C_{2h}, as the plane of the double bond in the latter compound is a plane σ_h.

The absence of any symmetry element implies that the compound is 'asymmetric' or of the point group with symmetry C_1. The symbol C_1 reminds us that it is always possible to choose any axis and effect a rotation of $2\pi/1$ around this axis, with the effect of bringing the molecule back into coincidence with itself. The 1-methyl-2-ethylcyclopropanes (Figure 4.1) belong to the group C_1.

Chirality and molecular symmetry

Symmetry operations can be grouped into two categories: the pure rotations C_n; and the symmetries related to a plane, centre or axis S_n, that is, generally speaking, the operations S_n $(n \geqslant 1)$ which are characteristic of the 'symmetry of reflection'.

In the 'symmetry of reflection' at least one operation of symmetry related to a plane is always involved.

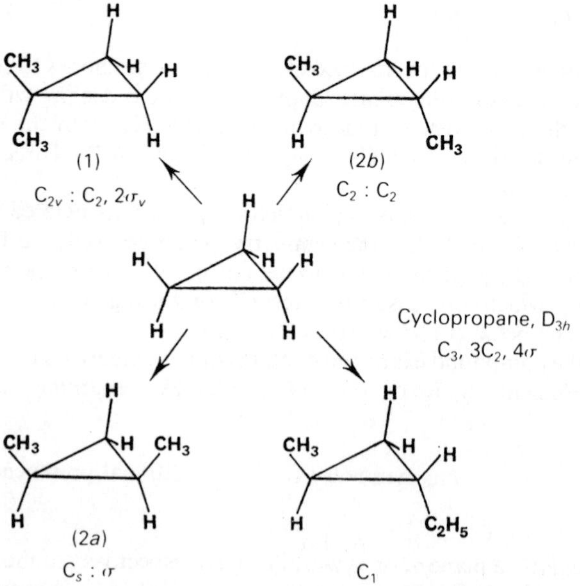

Figure 4.4 Reduction of the symmetry of cyclopropane by the introduction of substituents

A molecule, A, is chiral when its mirror image, B, is not superimposable on A. If A contains a plane of symmetry its image B will necessarily be equivalent to A. To verify this, one only needs to make the plane of molecular symmetry of A the mirror plane, so that the image, B, by definition coincides with A. Molecular chirality is incompatible with a plane of symmetry and, generally, with any element of symmetry of reflection.

In short, among the symmetry point groups, only the groups C_1, C_n and D_n are compatible with chirality. Compounds belonging to group C_1 are called asymmetric (total absence of symmetry elements) while molecules of the groups C_n and D_n are chiral and of axial symmetry.

The example of cyclopropane compounds (Figure 4.4) will again be used to exemplify the relations between chirality and symmetry elements.

Cyclopropane (D_{3h}) is very symmetrical, but the substitution of one or more hydrogen atoms by radicals reduces molecular symmetry and causes the disappearance of some or all of the symmetry elements. In Figure 4.4, the structures C_s and C_{2v} are achiral (having reflection symmetry), the cyclopropanes C_2 and C_1 are chiral.

The adjective 'asymmetric' should be used with care, as it literally implies the absence of symmetry. Chiral has a more general meaning, chirality describing the two situations below:

Chirality—
↗ Asymmetry
   (Absence of symmetry elements)

↘ Axial symmetry
   (Axes of symmetry, with the exception of improper axes S_n)

There is often a practical problem when the geometry of a molecule has been determined: deciding whether it is chiral, and therefore capable of enantiomerism.

(5)   (6)   Observation along C_3

One can solve the problem by two methods: either by constructing the mirror image molecule and verifying the identity or non-identity of the two structures, or by examining symmetry properties of the compound. To illustrate the latter method consider two examples. The tetraphosphine (5) has neither plane nor centre of symmetry, but it is not chiral because it contains an S_4 axis. The trioxalatoiridium ion $[Ir(C_2O_4)_3]^{3-}$ (6) contains an axis of symmetry of order 3 and three axes of symmetry of order 2. This complex, which belongs to the symmetry group D_3, is thus chiral; two enantiomers have actually been isolated by Délepine.

Experimental characterization of enantiomers

Two enantiomers have identical chemical and physical properties as long as the reagent or the reaction being considered possesses symmetry of reflection. It is, however, easy to demonstrate the difference in behaviour between enantiomers by using a chiral reagent or reaction. This is of fundamental importance and must be emphasized by several simple examples which are not all chemical.

It is difficult, but not impossible, to distinguish chemically between two enantiomers. Enzymatic reactions are cases where enantiomers react in a completely different manner. Amino acids of the D configuration are degraded in the presence of D-amino oxidase, an enzyme extracted from cobra venom (and which can be imitated by a chiral catalyst):

$$R—CH—CO_2H \xrightarrow[O_2, H_2O]{\text{D-amino oxidase}} \text{no reaction}$$

NH$_2$
L-enantiomer

$$R—CH—CO_2H \xrightarrow[O_2, H_2O]{\text{D-amino oxidase}} R—\overset{O}{\overset{\|}{C}}—CO_2H + H_2O_2 + NH_3$$

NH$_2$
D-enantiomer

Therefore, enzymatic tests for distinguishing one enantiomer from another can be carried out, based on the different reactivity of a reagent or chiral catalyst towards two enantiomers. To understand this phenomenon, compare a chemical reaction with the work of a craftsman. Consider a craftsman who is neither right-handed nor left-handed; he can carry out the same work with either hand if he has a symmetrical tool. However, if he is given a tool having a chiral handle he will only be able to work well with the hand which is adapted to the chirality of the handle.

Another analogous 'reaction' is the selection of right-hand gloves from a mixture of left and right gloves. This distinction can be made conveniently by trying to put each glove on the right hand, thus selecting all the right-hand gloves.

The biological activities of asymmetric chemical compounds are often linked to molecular chirality. Biological activity is thus a convenient method, in certain cases, for differentiating between two enantiomers. For example, D-asparagine:

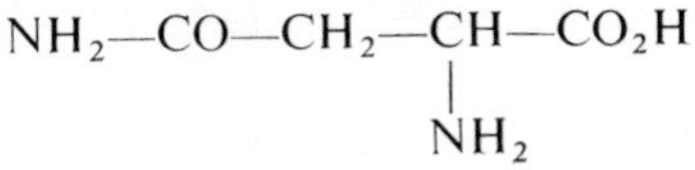

$$NH_2—CO—CH_2—\underset{\underset{NH_2}{|}}{CH}—CO_2H$$

is sweet, whereas its enantiomer, L-asparagine, is bitter. (—)-Methadone:

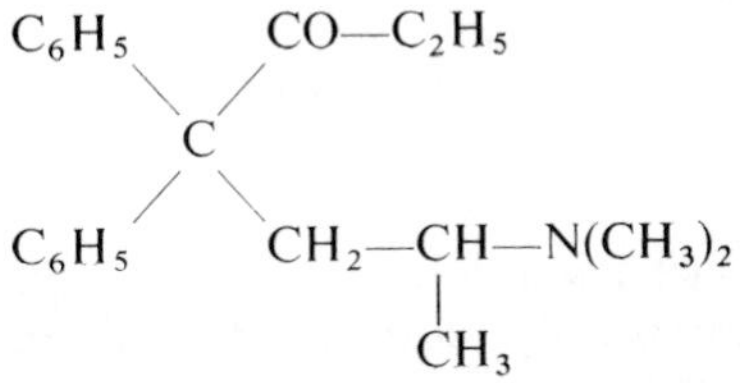

$$\underset{C_6H_5}{\overset{C_6H_5}{}}\diagdown\underset{\diagup}{\overset{\diagdown}{C}}\diagup\overset{CO—C_2H_5}{\underset{CH_2—\underset{\underset{CH_3}{|}}{CH}—N(CH_3)_2}{}}$$

is a strong analgesic, whereas its enantiomer is practically inactive. In examples of this type, theories have been put forward implying active chiral sites which can interact efficiently with only one enantiomer. If one assumes a specific mode of adsorption by three points (the rule of three points) it is obvious (Figure 4.5) that an asymmetric selection will operate (as in the testing of gloves); the correct enantiomer will be adsorbed, by establishing the necessary 'contact' for biological activity:

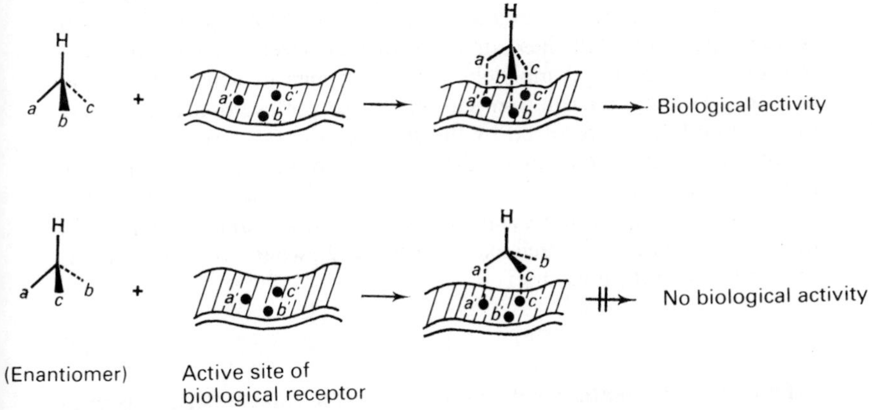

(Enantiomer)   Active site of
                     biological receptor

Figure 4.5

Among the physical methods used for distinguishing between two enantiomers, the measurement of optical rotatory power is undoubtedly the best (Figure 4.6). Two enantiomers rotate the plane of plane-polarized light in opposite directions and by an equal angle, in consequence of which the name optical isomerism is often used instead of enantiomerism.

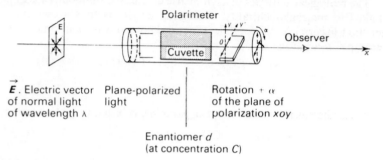

Figure 4.6

An enantiomer which effects a rotation α of the plane of polarization in the clockwise sense is called dextrorotatory (d). In the opposite case it is called levorotatory (l). It is conventional to characterize an optically active compound by its specific rotation $[\alpha]$, given by formula (1) (Biot's law):

$$[\alpha] = \frac{\alpha}{Cl} \tag{1}$$

The rotation α, measured polarimetrically, is found to be proportional to the concentration C (in $g\,cm^{-3}$) and to the length l (in dm) of the cuvette traversed by the polarized light. $[\alpha]$ is independent of l but varies with the wavelength λ, the solvent and (slightly) with the concentration C. It is thus necessary to state these three parameters as well as the temperature in order to define $[\alpha]$ completely. If the specimen of which one wishes to measure the specific rotation is liquid, the solvent is not necessary and, in equation (1), C is replaced by the specific gravity in $g\,cm^{-3}$.

One often wishes to compare the rotatory powers of analogous compounds with different molecular weights. The comparisons are more useful if one employs the molecular rotation $[\Phi]$ which take account of the molecular weight. $[\Phi]$ is defined by formula (2), where M represents the molecular weight (in grams):

$$[\Phi] = [\alpha] \times \frac{M}{100} \tag{2}$$

To identify an enantiomer it is necessary only to state the sign of its rotatory power (for given experimental conditions). For example, in acetic acid solution, with yellow light ($\lambda = 589$ nm), natural alanine $CH_3-CH-CO_2H$
$\qquad\qquad\qquad\qquad\qquad\qquad\qquad\qquad\qquad\qquad\quad |$
$\qquad\qquad\qquad\qquad\qquad\qquad\qquad\qquad\qquad\qquad NH_2$

is dextrorotatory and may be called d-alanine or $(+)$-alanine. Its enantiomer is l-alanine or $(-)$-alanine.

Racemates and optical purity

The synthesis of pure enantiomers, which occur naturally in living organisms, can be a difficult problem in the laboratory. This will be discussed later along with ways of approaching pure enantiomers. Experimentally, equimolecular mixtures of enantiomers are encountered very frequently and are called racemic mixtures. The rotation of a racemic mixture is zero due to compensation of rotations.

It is also common to encounter mixtures enriched in one of the enantiomers. These mixtures have rotations $[\alpha]$ dependent on their composition. If $[\alpha]_{max}$ is the specific rotation of one enantiomer, the fraction $[\alpha]/[\alpha]_{max} = P$ is called the optical purity. Knowing P enables one to calculate the enantiomeric composition of the mixture, d/l. On the hypothesis that the rotations produced by each enantiomer are additive it is easy to demonstrate the relationship (3):

$$P = \frac{d - l}{d + l} \tag{3}$$

where d and l represent the number of molecules of each enantiomer, and $(d-l)/(d+l)$ is called the enantiomeric purity.

The enantiomeric composition d/l can be calculated using formula (4):

$$d/l = \frac{1 + P}{1 - P} \tag{4}$$

When one wishes to calculate the d/l ratio precisely it can be useful to have methods other than those based on polarimetry available. There is the risk that optical purity may be different from enantiomeric purity, for example for compounds strongly associated by hydrogen bonds. Let the associations of each enantiomer be represented as $d\ldots(d)_n$ and $l\ldots(l)_n$ respectively. Then consider a mixture of enantiomers; new associations $l\ldots(d)_n$ and $d\ldots(l)_n$ appear and will make their own contribution to the rotation of the mixture. The relationship $P=(d-l)/(d+l)$ does not take account of such effects, and the calculation of the ratio d/l may thus be in error. The first example of the non-identity of enantiomeric purity and optical purity has recently been described by Horeau[1] for α-methyl-α-ethylsuccinic acid:

$$\begin{array}{c} CH_3 \\ \diagdown \\ C{-}CH_2CO_2H \\ \diagup \quad | \\ C_2H_5 \quad CO_2H \end{array}$$

[1] A. Horeau, *Tetrahedron Letters*, 1969, **36**, 3121.

Some distinctions between racemic compounds and enantiomers   In solids, simple physical properties often allow us to distinguish between a racemic mixture and its constituents. Very often a racemic mixture crystallizes in a crystallographic system different from that of the corresponding enantiomers, the crystalline lattice incorporating both enantiomers. A new compound has, therefore, been formed in the solid state and is called a racemate. When dissolved it decomposes, liberating its constituents. The racemate has a melting point different from that of each enantiomer. Sometimes crystallization takes place quite differently, one enantiomer only being present in each crystal. When the crystallization is finished the substance consists of as many right-hand crystals as left-hand crystals. This mixture is called a racemic conglomerate. A mixture of two crystalline types necessarily has a melting point different from each of the types, which again gives a means of distinguishing between the enantiomers and the racemic mixture, simply by examining the melting point.

Figure 4.7 symbolizes the crystallization of a racemic mixture, leading either to the racemate or to the conglomerate.

1. Crystallization of an optically pure compound:

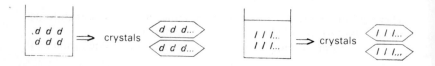

2. Crystallization of a racemic mixture:

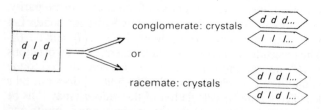

Figure 4.7

The relationship between the melting point of a solid and its enantiomeric composition has been studied extensively. Thermal analysis diagrams clearly reveal the racemate or conglomerate nature of the racemic mixture in the solid state.

The conglomerate is a true eutectic mixture, and its melting point is necessarily lower than that of each of its enantiomers. The melting point of the racemate may be either lower or higher than that of its constituents; the two possible situations are illustrated below.

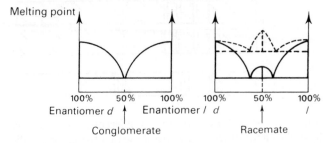

Two crystals which are not identical (because of the arrangement of the components in the crystalline lattice) may often be distinguished from one another by means of their physical properties. A racemate can be considered as a solid having its own characteristics, which differ from those of the solid phases prepared from pure enantiomers.

An infrared spectrum of the solid or a powder X-ray diffraction diagram (Debye–Scherrer diagram) can be a very effective method for distinguishing between a racemic mixture and an optically pure compound, provided that the racemic compound crystallizes in the form of a racemate. An example is that of Schlenk[1], concerning the study of the triglycerides which are encountered in natural fats.

$$R-\underset{\underset{O}{\|}}{C}-O-CH_2-\underset{\underset{\underset{COR''}{|}}{O}}{CH}-CH_2-O-\underset{\underset{O}{\|}}{C}-R'$$

The fatty acids which esterify glycerol are generally very similar; it is not enough to have $R \neq R'$ in order that the compound be optically active. One thus encounters a paradoxical situation: it is possible for a triglyceride to be 'optically pure', or, more exactly, enantiomerically pure, while being optically inactive. In these conditions, how can one know whether a given triglyceride is racemic or not? A simple solution to the problem involves preparing the same triglyceride, either in a racemic state or enantiomerically pure, in the laboratory, and comparing the X-ray diagrams of the various specimens. For example the 1-laurate, 2,3-dipalmitate of glycerol,

$$CH_3-(CH_2)_{10}-\underset{\underset{O}{\|}}{C}-O-CH_2-\overset{*}{\underset{\underset{\underset{O=C-(CH_2)_{14}-CH_3}{|}}{O}}{CH}}-CH_2-O-\underset{\underset{O}{\|}}{C}-(CH_2)_{14}-CH_3$$

[1] W. Schlenk, *Angewandte Chem. Intern.*, 1965, **4**, 139.

gives the following X-ray diagrams:

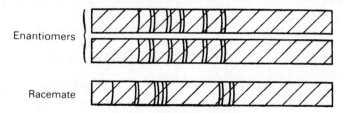

Enantiomers

Racemate

It is thus easy to use these patterns to differentiate the racemic compound from the enantiomers; they also show that the racemic solid exists as a racemate, otherwise the three patterns would be completely identical.

It is always useful to know whether a racemic compound crystallizes as a conglomerate or as a racemate. The spontaneous or mechanical resolution of the racemic compound can be carried out only in the case of a conglomerate. The resolution of a racemic compound is an operation which consists of separating the two optical enantiomers. It involves actually sorting the molecules:

$$
\begin{array}{c}
d \\
l \quad d \quad l \\
d \quad l \quad l \\
d
\end{array}
\xrightarrow[\text{resolution}]{}
\quad
\begin{array}{c}
\left.\begin{array}{c} d \quad d \\ d \quad d \end{array}\right\} \; 100\% \; d \\
\hline
\left.\begin{array}{c} l \quad l \\ l \quad l \end{array}\right\} \; 100\% \; l
\end{array}
$$

$$
\underbrace{\qquad\qquad}_{\begin{array}{c}50\% \; d \\ 50\% \; l\end{array}}
$$

This resolution can be carried out by chemical or biochemical methods. The simplest method consists of inducing crystallization of one enantiomer in a supersaturated solution of the racemic compound by seeding with a small crystal. If crystallization is completely controlled, one enantiomer will be separated in the crystalline form, for example the d-form, leaving the l-form in solution. In general, this operation is incompatible with the crystallization of a racemate; it is necessary that the d seed, for example, should surround itself with molecules identical to it, as is the case with racemic conglomerates. This process of resolution, described as preferential seeding, requires no additional optically active material except the first seed to initiate it. It is used on an industrial scale, for example in resolving racemic glutamic acid,

$$
\begin{array}{c}
HO_2C—CH_2—CH_2—CH—CO_2H. \\
\big| \\
NH_2
\end{array}
$$

Unfortunately, this method of resolution, whether spontaneous or initiatiated, remains limited in scope because of the small number of racemic compounds which crystallize in the state of conglomerates. According to recent statistics (1972) due to J. Jacques[1], there are at present 124 known cases of spontaneous resolution.

Chiral molecular structures

All molecules without the characteristic symmetry elements of reflection (axes S_n, or plane σ) are, by definition, chiral and can in principle give rise to optical isomerism or enantiomerism. It is usual to distinguish between the several causes of enantiomerism.

Asymmetric atoms

Asymmetric carbon   A carbon atom of sp^3 hybridization, substituted by four different groups, that is of type *Cabcd*, is not superimposable on its mirror image, because of the non-coplanarity of the four valencies:

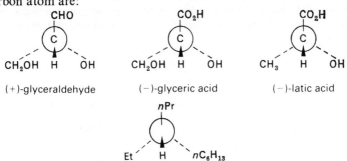

E   ⊐   Different from E

Irrespective of the symmetrical character of the four substituents, the elements of symmetry originally present in CH_4 (p. 85) have disappeared. For example, the plane *aCb* is not a plane of molecular symmetry. It has become conventional to call the central carbon atom of substances *Cabcd* 'asymmetric carbon' (absence of any symmetry element). Describing the carbon atom is obviously only a simplification; it is the whole molecular structure which is asymmetric, and which belongs to point group C_1.

Examples of molecules made chiral by the presence of an asymmetric carbon atom are:

CHO CO_2H CO_2H

C   C   C

CH_2OH H   OH CH_2OH H   OH CH_3 H   OH

(+)-glyceraldehyde   (−)-glyceric acid   (−)-latic acid

nPr

Et   H nC_6H_{13}

2-ethyldecane ≡ ethylpropylhexylmethane

[1] A. Collet, M. J. Brienne, J. Jacques, *Bull. Soc. Chim.*, 1972, 127.

The magnitude of the optical rotation which is associated with the presence of an asymmetric carbon atom is very variable, and depends basically on the nature of the substituents. The absence of a rotation does not necessarily signify that one is dealing with an achiral or racemic compound. One example of a compound which is enantiomerically pure but without any rotatory power was given on p. 97. Another example is ethyl propyl hexyl methane, prepared by Wynberg, which is without rotation for all the wavelengths accessible with a polarimeter. It is interesting to note, however, that the specific rotations of molecules made asymmetric by the substitution of one hydrogen atom by a deuterium atom have been measured. Monodeuterated phenylethane, $C_6H_5—CHD—CH_3$, has a specific rotation:

$$[\alpha]_D = 0° 30'.$$

It is important to remember that a molecule containing several asymmetric carbon atoms is not necessarily chiral. Molecular geometry may introduce an element of symmetry of reflection which renders the compound achiral (a meso compound): cis-1,2-dimethylcyclopropane (2a), (page 90), possesses a plane of symmetry perpendicular to the plane of the ring, and is thus without chirality.

Asymmetric heteroatoms   Many atoms other than carbon can play the role of an asymmetric atom.

Nitrogen is pyramidal in amines, but it is impossible to isolate optical enantiomers because the inversion of valencies at this atom is very rapid at ordinary temperatures:

Inhibited inversion of nitrogen in amines was first observed in 1956, when Prelog resolved 'Tröger's base', obtaining optically active forms. Tröger's base contains two asymmetric nitrogen atoms and belongs to the group C_2:

Tröger's base   (7)   (8)

In oxides such as (7) or in ammonium ions like (8) the inversion is similarly prevented, thus permitting the isolation of enantiomers. The presence of optical activity in the amine oxide (7) is a good demonstration that the semi-polar linkage $\overset{+}{N}-\overset{-}{O}$ conserves the tetrahedral character of nitrogen.

In complex (9) nitrogen is coordinated to cobalt and constitutes the only centre of asymmetry. This compound has been resolved.

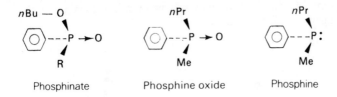

(9)

Sulphur gives rise to a whole group of stable chiral compounds. Optical activity has been demonstrated in suitably substituted sulphinates, sulphoxides and sulphonium cations:

Me —⬡---S⁺:   ⬡---S⁺:   HO — C — CH₂ ---S⊕:
       OEt   CO₂H Me   O   Me

Chiral sulphinate   Chiral sulphoxide   Chiral sulphonium cation

Phosphorus behaves as as an asymmetric atom in many derivatives; some organophosphorus compounds of which the optical enantiomers have been prepared are shown below.

nBu — O   nPr   nPr
⬡---P →O   ⬡---P → O   ⬡---P:
      R   Me   Me

Phosphinate   Phosphine oxide   Phosphine

nPr   Me
⬡---P⊕ — Me   ⬡---P = NR
   C₆H₅CH₂   nPr

Phosphonium cation   Phosphine imine

Optically active pentacoordinated compounds of phosphorus were prepared in 1966, clearly showing the non-planar character of phosphorus in these derivatives:

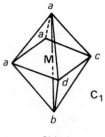

Silicon is capable of replacing carbon in many organic compounds and can similarly play the role of an asymmetric atom when the four substituents which are attached to it are all different.

This list of heteroatoms capable of acting as an asymmetric centre is far from complete; in particular, selenium, tellurium and arsenic should be mentioned. Metallic complexes constitute an interesting and very varied group, in which chiral molecules are found where the ligands are not necessarily asymmetric themselves. Only the stereochemistry of some octahedral complexes, in which the central metal can be Co, Ni, Pt, Cr, Rh, Ir, Fe, will be dealt with here. At the beginning of this century Werner demonstrated the structure of many transition-metal complexes. In contrast to the case of tetrahedral coordination (e.g. in compounds of carbon) many of the ligands may be identical in an octahedron, without necessarily preventing chirality in the structure. For example, a complex of the type Ma_3bcd may or may not be chiral, according to the relative arrangement of the ligands:

Chiral   Achiral
(abcd is a plane of symmetry)

When the ligand is bidentate (symbolized by a—a), the chiral octahedral complexes may have axial symmetry:

$$M(a - a)_2 \, bc \qquad\qquad M(a --- a)_2 \, b_2 \qquad\qquad M(a -- a)_3$$

The complexes $M(a$—$a)_3$ belong to the point symmetry group D_3; for example, trioxalatoiridium^{3-} which was discussed on page 92.

Molecular chirality in the absence of asymmetric atoms

The allenes are a group of which the optical activity was predicted by van't Hoff sixty years before the first optically active allene (10) was obtained. This allene possesses binary symmetry C_2, easily detectable using the projection which corresponds to observation of the molecule along the axis C=C=C:

(10)

C_1 asymmetric allenes are also known, for example (11). It is interesting that natural allenes have been isolated in an optically active form. Carbodiimides can be considered as nitrogen analogues of allenes, and the carbodiimide containing ferrocene nuclei (12) has recently been resolved:

(11)   (12)

The 4-substituted alkylidenecyclohexanes and related compounds are analogous to the allenes, one of the allenic double bonds being replaced by a ring; in consequence, the terminal groups lie in perpendicular planes. For example, the acid (13), when observed along the axis of its double bond, gives the projection:

(13)

Such compounds are chiral and have actually been obtained in an optically active form, as have the oxime (14) and the hydrazone (15).

(14)   (15)

Spiranes such as (16) or (17), having substituents at each end of the molecule, can be regarded as derived from an allene in which both the double bonds are replaced by rings:

Molecular chirality is typical of spiranes (16) and (17). The spirohydantoin (19), the borosalicylate (18) and the spirolactone (20) have also been obtained in optically active form. These compounds have no real analogy with the allenes.

The resolution of the spiranes (17) and (18) is a good demonstration of the non-planarity of the atoms of nitrogen and boron in these compounds. The spiranes have been classified among compounds without asymmetric carbon atoms; this grouping is convenient but to some extent arbitrary. In these compounds there is always an atom A which is tetra-coordinated, apparently of the type A*aabc*, and therefore not asymmetric. In fact, careful examination shows that this atom sometimes is asymmetric; this will be discussed on page 115.

(16)   (17)   (18)

(19)   CO$_2$H   (20)

Ortho-disubstituted biphenyls have a preferred conformation which is non-planar in consequence of the steric interaction between the groups in the *ortho* positions.

Planar conformation
The aromatic nuclei are
conjugated, but there is
some steric strain

Preferred nonplanar
conformation

The conditions for chirality in biphenyls are equivalent to those observed in the case of allenes. It is necessary that neither nucleus possesses a plane of symmetry passing through the other ring, and consequently that A ≠ B and A′ ≠ B′. The demonstration of biphenyl chirality by the optical activity of each enantiomer requires that the life of that enantiomer shall be of sufficient length. In contrast to allenes, it is necessary to consider an energy factor, related to the nature of the substituents placed in the *ortho* positions. In Figure 4.8 the variation in energy accompanying rotation of a phenyl ring around the bond joining the two rings is shown qualitatively.

The simple case, in which the same substituents are present in each ring, is given as an example. Two energy barriers to rotation can be distinguished, having symmetry C_{2v} and C_{2h} and separating two chiral conformations of symmetry C_2. The lower barrier, C_{2h}, represents the more probable transition state by which the interconversion between the C_2 enantiomers takes place. To avoid appreciable racemization at room temperature it is necessary that this lower rotational barrier should exceed 80–90 kJ mol^{-1}. In practice, examples are known of all the possible cases between very fugitive optical activity at ordinary temperature and complete optical stability even at high temperatures. The rate of rotation around the linkage between two rings can be controlled by the choice of substituents. The replacement of the *ortho*-hydrogen atoms in (21) by deuterium has made it possible to demonstrate a secondary isotope effect, due to the steric hindrance of deuterium being smaller; the deuterated biphenyl racemizes more rapidly than its analogue (21) containing hydrogen.

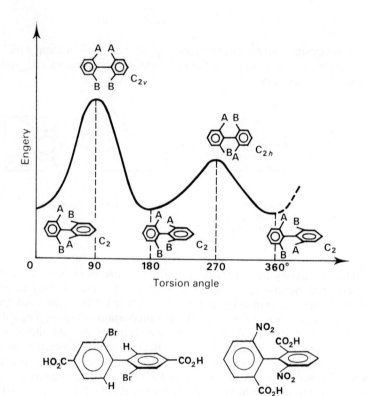

(21)

Easy racemization

(22)

Great optical stability

Figure 4.8

The hindered rotation characteristic of biphenyl isomerism is also found in related structures. For example, the tricyclic systems (23) and (24) have been obtained in optically active forms, the terminal rings constituting the elements of a biphenyl. They are not coplanar because of steric strain.

'Bridged' biphenyl   (23)   (24)

Atropoisomerism and conformational chirality

The demonstration of chirality in the biphenyl series was the first experimental indication that rotation around a single bond (in this case between two sp^2 hybridized carbon atoms) can be prevented completely in some instances. The name 'atropoisomerism' has been proposed to characterize the stereoisomers (enantiomers or diastereomers) which are the direct consequence of this hindrance of rotation. There are many cases of atropoisomerism related to biphenyl isomerism, with the consequent appearance of chirality. For example, the styrene (25) and the sulphonamide (26) have been resolved. In these compounds an aromatic ring is joined to a planar system (double bond or nitrogen atom respectively) which cannot become coplanar with the ring because of steric interactions:

(25)   (26)   (27)

Bonds in which rotation is hindered are shown as heavy lines

The substituted benzophenone (27) has recently been resolved, thus demonstrating simultaneously the non-planarity of the compound and the hindrance of rotation around the bonds attached to the carbonyl group.

Atropoisomerism, in principle, refers to the hindrance of rotation (torsional isomerism); however, the term is sometimes extended to cyclic molecules without asymmetric atoms but made resolvable because of the stability of a chiral conformation. The term 'conformational chirality' describes the situation equally well without specifying the route from a conformer to its enantiomer, which in fact requires a complex set of rotations and angular deformations. In Figure 4.9 several cases of conformational chirality of very different types are shown, in all of which steric strain prohibits the planar conformation. For example, hexahelicene cannot be planar as this would involve the superimposition of the atoms of the terminal rings. In consequence the system is deformed and takes up a helical conformation, which has a remarkable rotatory power ($[\alpha] = \pm 3600°$).

Trans-cyclo-octene is a very strained compound which has been resolved; so also have 'ansa' derivatives and paracyclophanes.

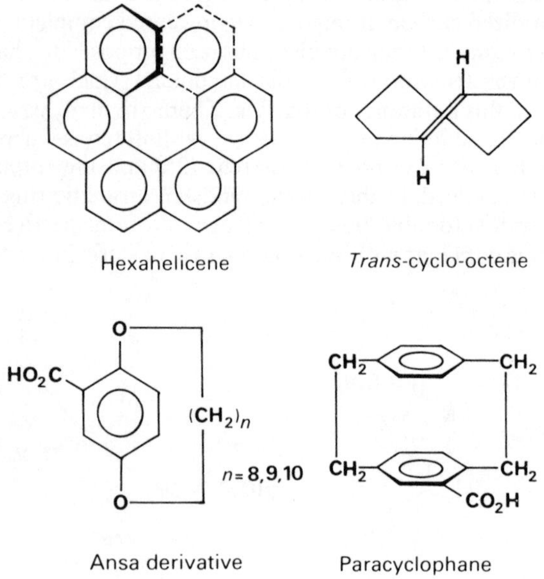

Hexahelicene   *Trans*-cyclo-octene

Ansa derivative   Paracyclophane

Figure 4.9

Absolute configuration

Two pure enantiomers are distinguishable from one another by physical or chemical properties (for example, by the sign of the optical rotation). An important question remains: to assign to one given isomer (for example, the dextrorotatory isomer) its correct spatial formula. There is a choice to make between two structures which are mirror images of one another; when this has been done, the absolute configuration of the enantiomer is known. For example, natural alanine, characterized by $[\alpha]_D^{27} = +32° 3'$ ($c = 0.6$ in acetic acid), is constituted entirely of molecules (28), to the exclusion of their enantiomers (29). Therefore (+)alanine has the absolute configuration represented in (28).

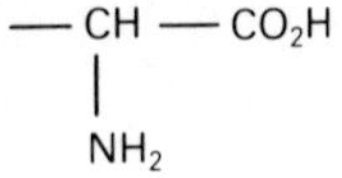

$$— CH — CO_2H$$
$$|$$
$$NH_2$$

Natural alanine

$[\alpha]_D = +32°3'\ (CH_3CO_2H)$

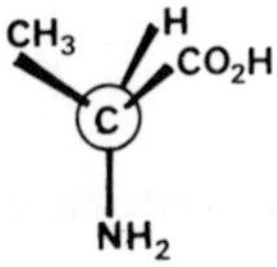

(28) 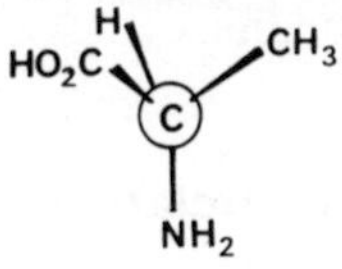   (29)

It is obvious that molecule (28) can be observed in various ways without its absolute configuration being changed. It should be verified that the representation (30) gives all the information necessary to allow the differentiation of (+)alanine (28) from its enantiomer (the rotational isomers about the various bonds do not need to be defined as they also occur in the enantiomer).

A perspective representation such as (30) or (31) (Figure 4.10) soon becomes difficult to draw if the molecule is complex; projections made with a certain number of conventions are usually preferable. Any projection allowing the reconstitution of the corresponding molecular model is a representation of absolute configuration. In a Fischer projection one observes the asymmetric carbon atom in such a way as to see the longest carbon chain vertically and bending backward, the oxygenated carbon atom being placed at the top. The four bonds from the asymmetric centre

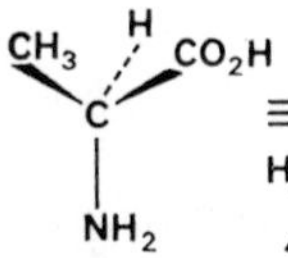

(30) 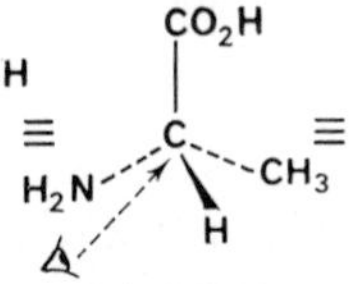   (31) 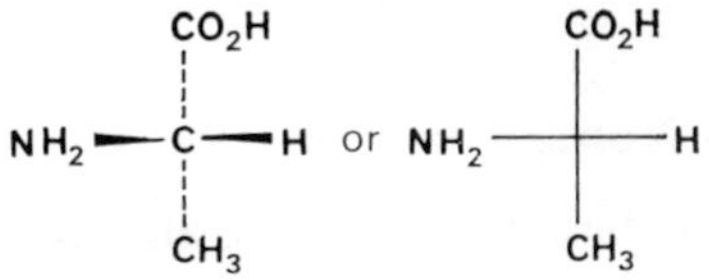   (32)

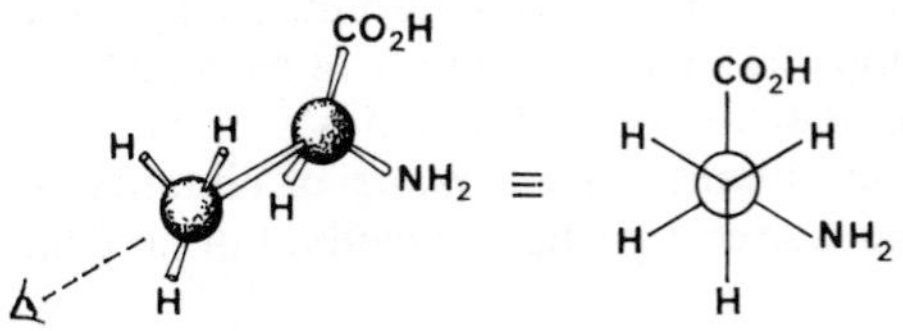

Figure 4.10   Various representations of natural alanine

are projected on to a plane passing through it. The two horizontal bonds are situated in front of the plane of the projection, thus (32) is the Fischer projection of (+)alanine. A Newman projection can often be convenient for showing the absolute configuration of a compound; (33) is a Newman projection of (+)alanine taken arbitrarily in one of its staggered conformations.

Nomenclature for absolute configuration

Instead of representing absolute configuration by a projection, it is simpler to give it a name. A descriptive term, generally a letter, allows us to state the absolute configuration, which amounts to a choice between two possibilities. The D, L nomenclature of Fischer applies to the Fischer projection. By convention (+)alanine will have the absolute configuration L, because the NH_2 group is placed on the left. The absolute configuration D characterizes the opposite arrangement (NH_2 on the right, H on the left). Generally, the D, L nomenclature can be applied without ambiguity whenever the horizontal bonds of the projection carry substituents H and X. Clearly the D, L nomenclature of absolute configuration implies a knowledge of the conventions which allow the writing of a Fischer projection, and thus the three-dimensional structure of the molecule.

It is important to avoid confusing d, l (dextrorotatory or levorotatory) and D, L (absolute configuration). A d compound may have either the absolute configuration D or the absolute configuration L; for example, (d)-L-alanine, (d)-D-glyceraldehyde. The Fischer nomenclature has proved particularly useful for natural products such as sugars or amino acids; however, it is not generally useful, and is being replaced by the R, S nomenclature of Cahn, Ingold and Prelog.

R, S Nomenclature

The establishment of a generally applicable system of nomenclature for absolute configuration has been found indispensable because of the great development of modern organic chemistry. In 1956 Cahn, Ingold and Prelog proposed rules of nomenclature for absolute configuration, applicable to every chiral compound with or without an asymmetric carbon atom. These authors consider three potential causes of chirality: a centre of chirality, an axis of chirality and a plane of chirality. Each of these chirality elements requires a particular method of nomenclature.

Elements of chirality
The classification of chirality based on symmetry elements requires a distinction among the symmetry groups C_1, C_n, and D_n. The systematic

nomenclature of enantiomers requires some new concepts. In a chiral molecule it is useful to consider the elements of 'dissymmetry' which are the cause of chirality, rather than the symmetry elements. For this purpose one assumes, sometimes arbitrarily, that chirality is the consequence of the destruction of a symmetry element in a precursor.

The idea of a centre of chirality embraces the notion of an asymmetric centre, but has a more general character. A centre of chirality is in most cases associated with a non-planar tetracoordinated atom, the four ligands all being different. An asymmetric carbon atom $Cabcd$ could be derived from an achiral precursor $Caabc$ where a ligand a is changed to a ligand d. This precursor (34) possesses planar symmetry. Its synthesis could also start from an achiral precursor $Caabb$ (35), which is more symmetrical as it contains two planes of symmetry and a C_2 axis at their intersection. The substitution in (34) of a radical a by the group d leads to the achiral distribution of four groups around the central carbon atom, which becomes the centre of chirality. The molecule belongs to the point group C_1, because all the elements of symmetry present in a precursor such as (34) or (35) have vanished.

© Centre of chirality

The centre of chirality linked to a chiral distribution of four radicals does not necessarily imply a molecular symmetry C_1 as in the classical case of asymmetric carbon $Cabcd$. For example, take an achiral compound $Caabb$ (35) and link each a group with a b group by equivalent rings. One thus obtains a spiran $abCab$ which retains only the binary axis of precursor (35). The spiran (18) is thus chiral with the symmetry point group C_2 (see also the compounds 19 and 20, p. 105). In this case the centre of chirality is the atom common to the two rings. Rules of nomenclature will be developed for the absolute configuration of such systems.

It is also theoretically possible to construct molecules having a centre of chirality of coordination number 4, and belonging to groups C_3 and D_2.

An axis of chirality is the consequence of a chiral distribution of substituents around an axis. Axial chirality can be imagined as a consequence of extending the tetrahedron of a precursor which may be chiral ($Cabcd$) or achiral ($Caabc$, $Caabb$). The imaginary extension leads to a diminution in molecular symmetry with the ultimate disappearance of symmetry of reflection. For example, the axial chirality for structures $C(ab)(ac)$:

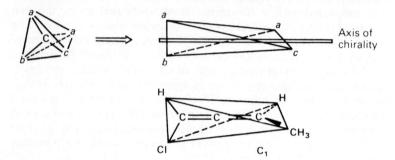

Axial chirality can easily lead to structures $C(ab)(ab)$ (for example, the allene (10) or the biphenyl (22), of symmetry C_2).

A plane of chirality is the result of the disappearance of a plane of symmetry through one or more modifications in the molecule which is thus rendered chiral. The paracyclophane (36) becomes chiral by carboxylation; the original plane of symmetry σ is called the chiral plane of the acid (37).

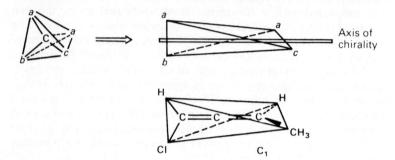

(36)   (37)

For convenience one chooses arbitrarily the plane of symmetry σ in (36) which contains the largest number of atoms, and one concludes that in the acid (37) there is a chiral distribution of atoms around this plane.

It is useful to consider the elements of chirality in a molecule; these elements of chirality emphasize the disappearance of elements of symmetry which leads to the appearance of chirality. This analysis will facilitate the nomenclature of absolute configuration, but does not replace the use of symmetry groups in the description of molecular geometry.

R, S nomenclature around a centre of chirality

Taking as an example an asymmetric atom A surrounded by four ligands a, b, c, d, the method of nomenclature requires several operations. One first of all arranges these groups in decreasing 'priority' (assume $a > b > c > d$); one then imagines that a, b and c are placed on a steering wheel of which the axis is A—d. If from a position above the wheel one observes a clockwise arrangement of the sequence a, b, c, the absolute configuration is called R (rectus) whereas in the opposite case it is called S (sinister):

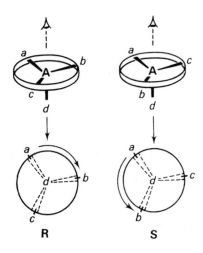

The criterion for the classification of the groups, that is their relative priority, is the comparison of the atomic numbers of the atoms directly bonded to the asymmetric atom; for example, $I > Br > Cl > F$. For substituents attached by a carbon atom, this given no definite order. The comparison between any two groups is therefore continued along the carbon chain as far as the end of the similarity. Thus, $-CH_2CH_3$ has priority over $-CH_3$, because the second carbon of the ethyl radical has priority over the corresponding hydrogen atom in the methyl group. When the chains are branched one continues the test by choosing the branching order with the maximum priority. The determination of 'priority' can be exemplified in detail in the following manner (see overleaf).

If one of the radicals contains a double bond, it is treated as though the atoms had the same degree of coordination that they would have in a saturated compound, e.g. with coordination number 4 for carbon. A double bond is imagined to be opened, and the addition, at each end, of the opposite atom is postulated. The vinyl group $-CH=CH_2$, for

$$- CH_2 - \overset{\overset{\displaystyle CH_3}{|}}{C}(CH_2OH)_2 \quad < \quad - CH_2 - \overset{\overset{\displaystyle CH_3}{|}}{\underset{\underset{\displaystyle CH_3}{|}}{C}} - CH_2Cl$$

Consideration of C_1 and test I: indefinite
Consideration of C_2 and test II: indefinite
Consideration of C_3 and test III: Cl>0, therefore $-CH_2-\overset{\overset{\displaystyle CH_3}{|}}{\underset{\underset{\displaystyle CH_3}{|}}{C}}-CH_2Cl$ has priority

example, is written   The two supplementary carbon atoms

are brought up to the valence of 4 by saturation with phantom atoms o of atomic number 0 (H, $Z = 1$, has priority over o, $Z = 0$). With this con-

vention the vinyl group can be represented as

It is, therefore, easy to see why $-CH=CH_2$ has precedence over $-CH_3$ and $-CH(CH_3)_2$. An analogous treatment leads to C=O being written in the form

The classification for the most common substituents is, therefore, as follows:

$$C_6H_5 > -\underset{\underset{CH_3}{|}}{\overset{\overset{CH_3}{|}}{C}}-CH_3 > -\underset{\underset{CH_3}{}}{\overset{\overset{CH_3}{}}{CH}} \quad > -CH_2-CH_3 > -CH_3$$

$$OH > NH_2 > CO_2H > CHO > CH_2OH > C_6H_5$$

These rules may now be applied for the naming of natural alanine:

In the case of a centre of chirality due to the presence of an isotope, it is the heavier isotope which has priority:

Absolute configuration R
(as D ⁄ H)

A centre of chirality present in certain spiranes (such as the spirohydantoin (19)) poses a problem of nomenclature because of the impossibility of comparing groups which are chemically and stereochemically equivalent. In the spirohydantoin (19) the two NH radicals bonded to the chiral centre have priority over the two carbonyl functions. Ambiguity

(19)   S configuration

is avoided by arbitrarily choosing one NH radical, which will be called NH$_{(1)}$ and to which priority is given over NH$_{(2)}$. The carbonyl group present in the ring containing NH$_{(1)}$ is called CO$_{(1)}$ and has priority over CO$_{(2)}$. With these conventions one finds an absolute configuration S for the spiran, a configuration which is independent of the original choice of the nitrogen atom.

R, S nomenclature in axial chirality

Starting with the deformed tetrahedron characteristic of axial chirality, the rule of the steering wheel is applied to the four groups at the corners of this tetrahedron. The following rule is used to classify the substituents. The tetrahedron is observed along the axis of chirality, and the two nearest groups are given priority over the groups placed at the far end of the axis (*see* Figure 4.11). It can be easily confirmed that the absolute configuration S of the dichloroallene illustrated is independent of the end of the axis of chirality which has been chosen to give priority to two of the substituents, relative to the other two.

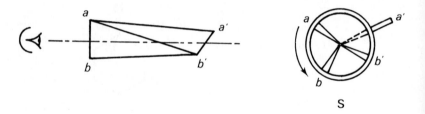

S

1. $(a,b) \sim (a',b')$ according to the rule of axial chirality

2. $\begin{array}{cc} a & \sim b \\ a' & \sim b' \end{array}$ (classified according to the general rules)

$a \sim b \sim a' \sim b'$

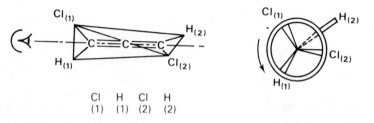

| Cl | H | Cl | H |
| (1) | (1) | (2) | (2) |

S configuration

Figure 4.11

Diastereoisomerism

Definition   Two stereoisomers which are not mirror images of one another with respect to a plane are described as diastereoisomers. In the introduction to this chapter the fundamental distinction between enantiomerism and diastereoisomerism was given: in two enantiomers the distances between nonbonded atoms are identical, whereas two diastereoisomers always have a certain number of distances between nonbonded atoms which are not identical. It follows that any molecular deformation of a compound, whether chiral or achiral, produces a diastereoisomeric structure (which is observable to the extent that its lifetime allows experimental study). An example outside chemistry is that of two corkscrews, A and A', having opposed screw threads. They are enantiomers of one another. If one of them is deformed (B) it becomes a diastereoisomer of the normal corkscrews A and A'.

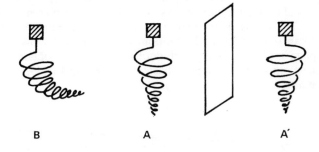

B   A   A'

This example helps in understanding the methods which can differentiate between stereoisomers. A symmetrical agent is incapable of distinguishing between the corkscrews A and A'; a cork, for example, is pierced with equal ease by A or A'. It is thus necessary to involve an external chiral agent (for example, a sheath having the form A, in which A' could not be inserted) in order to distinguish between enantiomers. This was discussed on page 92. On the other hand, a distinction between two diastereoisomers can be made without difficulty by a symmetrical reagent. The corkscrews A and B behave quite differently toward the cork, A can pierce the cork, B cannot. If one considers diastereoisomeric organic compounds, one comes to the same conclusion. Diastereoisomers have steric properties so different that they can be characterized experimentally without the use of a chiral test. In principle, the physical and chemical properties of diastereoisomers, such as the melting or boiling points, chromatographic behaviour, oxidation or reduction reactions, are different.

Two types of diastereoisomers can be distinguished using an arbitrary but convenient classification. Diastereoisomerism of torsion results from hindred rotation around one or more bonds. Intrinsic diastereoisomerism is independent of torsional isomerism; it is a form of stereoisomerism which arises when the molecule contains several asymmetric atoms. When there are different ways of combining the absolute configurations of asymmetric centres, each combination corresponds to one stereoisomer.

Diastereoisomerism due to torsion

Stereoisomerism due to torsion is observed when rotation around a bond is not completely free. The energetically disfavoured conformations constitute rotational barriers separating conformations of minimal energy. These last can be described as *rotamers*, and it has been seen, for example, that biphenyls substituted in the *ortho* positions with large groups exist in the form of rotameric enantiomers, of which one is represented by the structure (22). In certain terphenyls the hindrance to rotation leads to the simultaneous existence of enantiomerism and diastereoisomerism. The terphenyls (38), (39) and (40) (Figure 4.12) have been obtained pure: (38) is an enantiomer of (39) and a diastereoisomer of (40); (40) is achiral, as it contains a centre of symmetry.

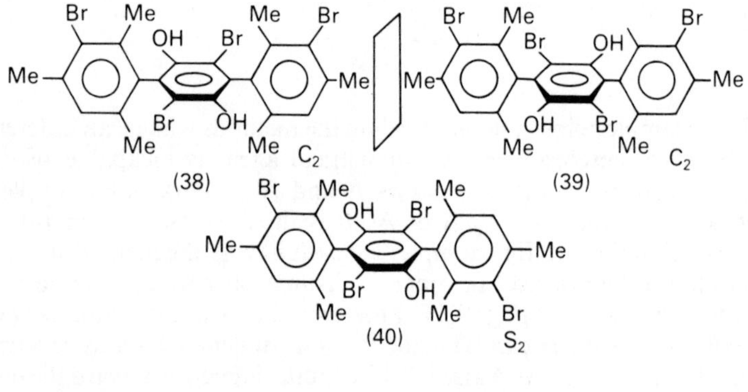

Figure 4.12

When the rotational barrier is too low to allow the isolation of diastereoisomers, these may sometimes be demonstrated by spectroscopic methods. For example, butadiene is a mixture of *s-cis* and *s-trans* conformations (*s* emphasizes that the torsion concerns a single bond); the terms *cisoid* and *transoid* are often used.

s-cis or cisoid   s-trans or transoid

It is interesting to note that it is the *s-cis* butadiene, the less abundant species at equilibrium, which reacts in cycloaddition reactions such as the diene synthesis.

The hindrance to rotation around a single bond increases with any increase in its double bond character; thus the C—N linkage of amides possesses a partial double bond character as a result of conjugation between the carbonyl group and the nitrogen atom. Torsional diastereoisomers are very often observed in amides of the type R—CO—N(R')(R'') as a result of the relative stability of the two planar conformations (*a*) and (*b*).

(a)   (b)   (41)   (42)

The n.m.r. spectrum of such an amide consists of the spectra of the diastereoisomers superimposed, which indicates that the interconversion is slow on the n.m.r. timescale. In some special cases it is possible to isolate, at room temperature, two amides which are torsional diastereoisomers, for example (41) and (42).

A double bond can be considered as an extreme case of hindered rotation, which brings geometrical isomerism about a double bond within the concept of torsional diastereoisomerism. This convention is convenient despite its arbitrariness; interconversion between geometrical isomers which exist because of differing configurations about certain C=N double bonds occurs as a result of rotation.

cis   trans

Geometrical isomers are differentiated by the prefixes *cis* or *trans* (or *syn/anti* when the isomerism is about C=N). This nomenclature sometimes causes difficulties. Its arbitrary character has been avoided by the Z, E nomenclature which has recently been proposed. At each end of the double bond we classify the substituents according to decreasing priority (using the rules of Cahn, Ingold and Prelog). The Z configuration (from the German 'Zusammen') is characterized by a *cis* relationship between the groups of highest priority; with a *trans* relationship the configuration is said to be E (from 'Entgegen'). The diacid (43) possesses an E configuration because the two groups of highest priority are *trans* to each other. The priority of the groups is established without difficulty:

$$Cl > CO_2H; \qquad CO_2H > H$$

cis	*trans*	*syn*	*anti*
E	Z	E	Z

The number of diastereoisomers increases rapidly with the possibilities of torsion in a molecule. For example, a diene such as piperylene:

$$CH_3-CH=CH-CH=CH_2$$

gives rise to four diastereoisomers:

cis s-cis	*cis s-trans*
trans s-cis	*trans s-trans*

Generally, when n is the number of localities for torsional isomerism in the molecule, one expects to find 2^n diastereoisomers. This total represents a maximum, because if the molecule has some degree of symmetry, some of the diastereoisomers will be identical. In 1,4-dimethylbutadiene, $CH_3-CH=CH-CH=CH-CH_3$, there are two positions possible for geometrical isomerism around a double bond, and one of torsional isomerism around the single bond; thus $n=3$, and theoretically $2^3 = 8$ diastereoisomers are possible. In fact there are only six diastereoisomers because of the equivalence of the double bonds, the isomer *cis s-cis trans* being identical to the isomer *trans s-cis cis* (Figure 4.13).

A chemist interested in synthesis considers only configurational isomers, the existence of which is independent of the *s-cis* or *s-trans* conformations; in the present case, there are only three defined chemical

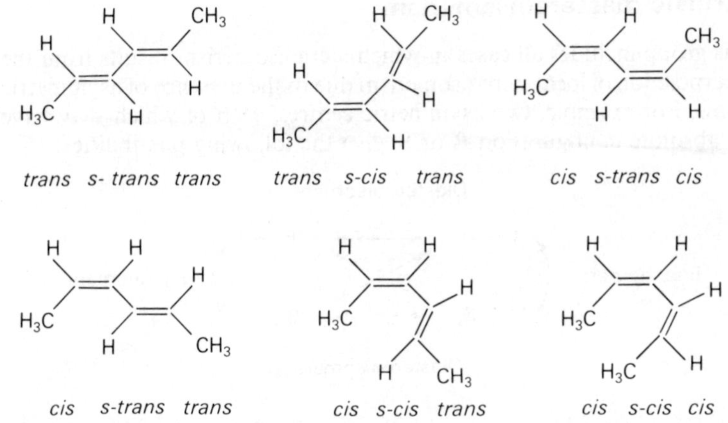

Figure 4.13

species at ordinary temperatures, the *cis-cis*, *trans-trans* and *cis-trans* diastereoisomers.

Cyclic systems

Cyclic systems in which the ring can adopt several non-equivalent conformations will be considered separately. A monosubstituted cyclohexane is a mixture of two conformations in equilibrium; in one the substituent R is equatorial, in the other it is axial. It is important to take account of the presence of these two diastereoisomers in the physicochemical study of such a compound. The isolation of the two conformational isomers of chlorocyclohexane has recently been achieved by Jensen[1]. As described on page 60, he separated the diastereoisomers (44) and (45) by crystallization at $-150°C$; each was quite stable at that temperature.

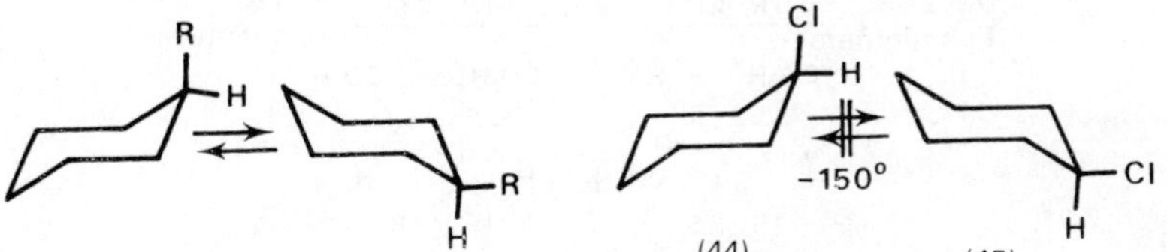

The separation of conformational isomers is still a very exceptional event, but n.m.r. study of equilibrium mixtures is now routine, and allow the evaluation of their relative stability.

[1] F. R. Jensen and C. H. Bushweller, *J. Amer. Chem. Soc.*, 1961, **91**, 3223.

Intrinsic diastereoisomerism

This group includes all cases in which stereoisomerism results from the superposition of local stereoisomerism due to the presence of asymmetric atoms. For example, two asymmetric centres, each of which may have the absolute configuration R or S, give the following possibilities:

Diastereoisomers

$$\text{Enantiomers} \left(\begin{array}{ccc} R - R & \longleftrightarrow & R - S \\ & \diagdown\diagup & \\ & \diagup\diagdown & \\ S - S & \longleftrightarrow & S - R \end{array} \right) \text{Enantiomers}$$

Diastereoisomers

R—R is diastereoisomeric with R—S or S—R, and enantiomeric with S—S. Generally, if there are n sources of chirality, 2^n stereoisomers should exist, forming 2^{n-1} pairs of enantiomers; any one isomer has $2^n - 2$ diastereoisomers[1]. Figure 4.1 (p. 81) shows the relationships which exist among several stereoisomers.

The preceding calculation does not involve molecular conformation, but treats stereoisomerism purely as configurational isomerism. To make an exact prediction, however, it is necessary to examine models to look for a possible symmetry element (a plane, a centre or an axis S_n) which may reduce a pair of enantiomers to a single meso compound.

Consider, for example, the cyclopropane-1,2-dicarboxylic acids. The diacid 1R—2S (*cis*) is identical with diacid 1S—2R, because of the plane of symmetry which prevents optical isomerism, so that there are three stereoisomers:

trans   *cis*
Enantiomers   1R 2R ___ 1R 2S   Identical
              1S 2S   1S 2R

Now consider the isomeric tartaric acids:

$$\text{HO}_2\text{C—CHOH—CHOH—CO}_2\text{H}.$$

[1] Two stereoisomers which differ in the absolute configuration of a single asymmetric centre are called *epimers*.

The molecular geometry is not as clearly defined as in the previous example, as rotational isomerism is possible. Neglecting this, one predicts four diastereoisomers ($R-R, S-S, R-S, S-R$). Whatever the molecular conformation of the isomers 2R—3R and 2S—3S, it is impossible to find a symmetry element which rules out chirality. On the other hand, the diastereoisomer of the configuration 2R—3S (or 2S—3R) can adopt a conformation which possesses a plane of symmetry σ, and as the interconversion of the various rotamers is rapid at ordinary temperature this acid is non-resolvable and is called the *meso*-acid:

2R 3R	2S 3S	2R 3S $\equiv$ 2S 3R	
Tartaric acids (enantiomers)		Mesotartaric acid	Chiral conformation of mesotartaric acid

It is interesting to note that *meso*tartaric acid possesses chiral conformations; one of these is shown above in the Newman projection. The racemization of such a chiral rotamer occurs by rotation around the central single bond.

The principle of nomenclature in intrinsic diastereoisomerism is simple; one enumerates the absolute configuration of all the asymmetric centres successively (for example, tartaric acid 2R—3R).

One particular nomenclature is frequently used for compounds containing two asymmetric carbon atoms. Consider two diastereoiosmeric sugars, erythrose and threose, with a planar formula

$$CHO-(CHOH)_2-CH_2OH,$$

and represented arbitrarily in an eclipsed conformation:

Erythrose
(2R,3R), *erythro*

Threose
(2S,3R), *threo*

The isomer 2R—3R is called *erythro*, the isomer 2S—3R is called *threo*. The designation of *threo* or *erythro* defines the relative configuration between two asymmetric centres. The configurations 2S—3R or 2R—3S characterize a relative stereochemistry *threo*; analogously the cyclopropane dicarboxylic acids 1R—2R or 1S—2S all have the *trans* relative stereochemistry. *Cis/trans* is used in preference to *erythro/threo* for cyclic systems containing two asymmetric carbon atoms. There is no rigorous definition of the relative stereochemistry *erythro* or *threo*. One usually compares two diastereoisomers which are assumed to take up an eclipsed conformation. The diastereoisomer which shows the maximum of similar or identical groups opposite each other is called *erythro*. This coincidence between analogous groups is better provided in erythrose (*erythro*-isomer): in threose (*threo*-isomer), the OH groups eclipse hydrogen atoms.

In this discussion of intrinsic asymmetry, only stable asymmetric centres have been considered. In some cases one of the asymmetric atoms inverts more or less easily, which makes it difficult to isolate the isomers. The nitrogen atom inverts rapidly, and therefore one cannot hope to prepare diastereoisomers of a structural isomer containing an asymmetric carbon atom and an asymmetric nitrogen atom. In some special cases it is possible to isolate the 'invertomers' of the nitrogen atom if this atom is part of a small ring. Thus the *cis/trans* isomers of some substituted aziridines have been successfully separated by Felix and Eschenmoser.

Prochirality

The concept of stereoisomerism implies the idea of comparison: in order to decide if a structural isomer gives rise to stereoisomerism a chemist will compare the various structures which can be constructed using the same formula. The non-identity of two models determines the existence of stereoisomers, and is shown by an 'external' comparison.

It is useful to examine a particular molecule in order to attempt to simplify the description of its structure with the aid of its symmetry properties, which are shown by an 'internal' comparison of the various parts of the molecule.

(1) Consider a molecule such as (46), having an axis of symmetry of order 2. A rotation of 180° brings CH_3 (*a*) into superposition with CH_3 (*b*) and vice versa, CO_2H (*a'*) places itself on CO_2H (*b'*) and CO_2H (*b'*) on CO_2H (*a'*). In this substituted cyclopropane, therefore, there is a geometrical equivalence between a certain number of the atoms or radicals. In particular, CO_2H (*a'*) is equivalent to CO_2H (*b'*); these two groups

necessarily have the same physicochemical properties. Any reagent (chiral or achiral) must react with the same speed at the sites a' or b', giving the same reaction product; for example, monoesterification at a' or b' leads to the acid esters (47) and (48), which are identical (one is converted to the other by a rotation C_2).

Physical properties are closely related to molecular symmetry. The n.m.r. spectrum of (46) reveals a single signal for the equivalent methyl groups a and b. However, in (47) the equivalence between the methyl groups has disappeared, each methyl being characterized by its own signal.

(2) Now consider a molecule having a plane of symmetry as its only symmetry element. The structure can be separated into two halves, each the mirror image of the other with respect to the plane of symmetry. The physicochemical properties of the two regions, which are called *enantiotopic*, are closely analogous but not strictly equivalent.

Methyl methylethylmalonate (49) is such a case. The two ester functions are placed on either side of a plane of symmetry and they thus have an enantiotopic character. The partial saponification affecting CO_2CH_3 (a) gives the monoester (51); the partial saponification affecting group (b) leads to (50), its enantiomer.

Partial saponification with sodium hydroxide will involve, with equal probability, reactions at sites (a) or (b), and the racemic mixture (50 + 51) will thus be obtained. It is, however, possible to imagine a hydrolysis, effected with an optically active base, which would take place specifically at site (a) or (b). A chiral reagent can distinguish between two enantiomers, and therefore also between the enantiotopic regions (a) and (b). This will be discussed in greater detail later. The n.m.r. spectrum does not distinguish between enantiomers; the same is true for enantiotopic substituents. In consequence, the methyl groups (a) and (b) will appear as a single signal.

(3) Compounds are often encountered which contain atoms or radicals which are chemically equivalent, but have physicochemical properties very different from each other. If the two substituents to be compared are situated in different chemical environments it is easy to understand that they will behave differently, for example, in methyl acetate,

CH_3—C—OCH_3 the two methyl groups are easily distinguished in the
$\overset{\|}{O}$

n.m.r. spectrum.

There is also the particular case where the two groups to be compared are situated on the same carbon atom. The word enantiotopic has been used to describe substituents when the plane of symmetry goes through the carbon atom which carries them. Now consider the example of a chiral molecule, such as (52), which is without a plane of symmetry.

Examination of the molecular model shows clearly the non-equivalence of groups CO_2CH_3 (a) and (b), which have a quite different steric environment ((b) is cis with respect to a vicinal hydrogen atom). The partial saponification of the ester function (a) leads to (54), which is a diastereoisomer of (53) obtained by a saponification at site (b). For this reason, the two substituents CO_2CH_3 are called *diastereotopic*. When a base such as sodium hydroxide is used, the functional groups (a) and (b) are not attacked at the same speed; (b) is hydrolysed more rapidly. In the n.m.r. spectrum the methyl groups (a) and (b) give separate signals.

Figure 4.14 summarizes the preceding discussion.

If the reaction $RR'CX_2 \rightarrow RR'CXY$ is considered in a general sense, it can be seen that when $R \neq R'$ a centre of chirality is created. Hanson has proposed the name '*prochiral* carbon' for any carbon atom of the type $(RR')C(X_2)$ containing a pair of identical ligands X. A prochiral carbon atom is present in structures (II)[1] and (III) (Figure 4.14), which is transformed into an asymmetric carbon atom, $(RR')C(XY)$, by a stereospecific reaction with one of the ligands X, in an asymmetric synthesis. Some examples of asymmetric syntheses are given in Chapter 5; here only the rules which make it possible to name each of the two ligands X will be considered.

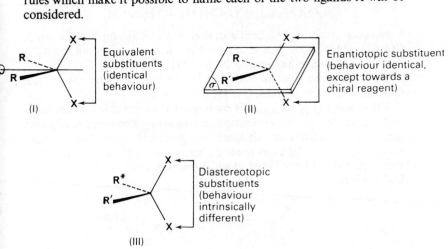

Figure 4.14

Nomenclature of prochirality

The need for a nomenclature of prochirality is clear since it was necessary to call the groups CO_2CH_3 of compounds (49) and (52) (a) and (b). In these two compounds ((II) and (III), Figure 4.14), a careful observer can always pick out one of the two X substituents without error. (III) contains an internal standard, the chiral group R*, which makes this observation easy. Thus, in (52) the functional group CO_2CH_3 (b) is identified without difficulty because of its *cis* relationship relative to the hydrogen atom of the neighbouring asymmetric carbon atom. In acyclic compounds this identification first requires the selection of an arbitrary conformation.

[1] It is sometimes wrongly called *meso*-carbon because of the presence of a plane of symmetry.

With the prochiral compounds (II) the difference between the two ligands X requires the aid of an external chiral agent, the observer for example.

An observer with his feet on R', and his head at R, and looking toward the prochiral carbon atom, will always see on his left the same substituent X (that which is below the plane σ of (II)).

Hanson has suggested calling the two X substituents *pro* R or *pro* S, enantiotopic (II) or diastereotopic (III). The convention is as follows: the group *pro* R is that which leads to a compound of absolute configuration R when it is accorded priority over the other group in the sequential rule of Cahn, Ingold and Prelog. For example, this convention will be used to name the diester (49). If CO_2CH_3 (*a*) is arbitrarily given priority over CO_2CH_3 (*b*), the ranking of the substituents will be

$$CO_2CH_3 \, (a) > CO_2CH_3 \, (b) > C_2H_5 > CH_3,$$

a sequence requiring the configuration S. CO_2CH_3 (*a*) is thus *pro* S, CO_2CH_3 (*b*) is *pro* R. The assignment *pro* R and *pro* S is unchanged if one chooses to give higher priority to CO_2CH_3 (*b*).

With the same rule we find that in the cyclopropane diester (52) CO_2CH_3 (*a*) is *pro* R and CO_2CH_3 (*b*) is *pro* S.

The concept of prochirality is particularly important for biochemists, since enzymatic systems are capable of interacting stereospecifically with one of two X substituents situated on a prochiral centre. For example, some enzymatic reductions require a coenzyme derived from nicotinamide (partial formula (55)), which donates hydrogen in the form of hydride ion:

(55)

According to which enzyme is involved it is either the *pro* R or the *pro* S hydrogen atom which will be transferred.

Generalization of the concept of prochirality

The formation of an asymmetric carbon atom starting with a prochiral carbon atom $RR'CX_2$ is not encountered frequently in the laboratory. The majority of asymmetric syntheses are effected starting with a trigonal carbon atom of the type $\begin{array}{c} R \\ \diagdown \\ \diagup \\ R' \end{array} C{=}A$, where A represents O, NH, CH_2, etc.

Consider, for example, the reduction of pyruvic acid, $\begin{array}{c} CH_3 \\ \diagdown \\ C=O, \text{ to} \\ \diagup \\ HO_2C \end{array}$

lactic acid. The addition of hydrogen on face 1 leads to S-lactic acid, addition on face 2 gives R-lactic acid (Figure 4.15):

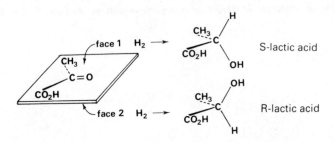

Figure 4.15

The two sides of the carbonyl group play the role of the ligands X on a prochiral carbon atom $RR'CX_2$. In the example in Figure 4.15 they have an enantiotopic character, and become diastereotopic if a source of chirality is present in the molecule.

To name the two non-equivalent sides of a trigonal carbon atom $Cabc$ the following convention is used:

(1) The three groups are ranked according to the sequential rules of Cahn, Ingold and Prelog.

(2) The face to be named is observed so that the three substituents are arranged in a circle. If the sequence $a>b>c$ defines a clockwise sense of rotation the face observed is called *re* (derived from rectus); in the opposite case the face is called *si* (derived from sinister).

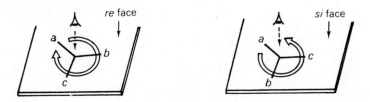

Face 1 of pyruvic acid ($O>CO_2H>CH_3$) shown in Figure 4.15 is *re*, face 2 is *si*. The reduction of pyruvic acid to S-lactic acid requires an attack on the *re* face.

Bibliography on stereoisomerism

E. L. Eliel, *Stereochemistry of Carbon Compounds*, McGraw-Hill, 1962.

E. L. Eliel, *Elements of Stereochemistry*, Wiley, 1969.

H. Kagan, Isomérie stérique, *Unichimie*, No. 2, 1971, 11.

K. Mislow, *Introduction to Stereochemistry*, W. A. Benjamin, 1965.

G. Natta and M. Farina, *Stereochemistry*, Longmans, 1972.

M. Orchin and H. Jaffé, Symmetry, Point Groups and Character Tables, *J. Chem. Ed.*, 1970, **47**, 246, 372, 510.

J.-L. Pierre, *Principes de stéréochimie organique statique*, Armand Colin, 1971.

5

Dynamic stereochemistry

Introduction

A knowledge of stereochemistry and molecular conformations would only be of limited interest to the chemist if he did not know how to use this information for predicting or explaining the chemical behaviour of compounds. Dynamic stereochemistry deals with the chemical properties of molecules in relation to the stereochemistry of the reactants and reaction products.

It is especially important to understand the stereochemical development of reactions under defined conditions, in order to control the stereochemistry of the various stages in an organic synthesis. The rational 'made to measure' synthesis of an organic compound is central to the practice of organic chemistry. Physical, therapeutic and alimentary properties are all closely involved in molecular stereochemistry. Consider, for example, the elasticity of rubber and the rigidity of gutta-percha, polymers which differ only in the *cis/trans* stereochemistry of their double bonds; or the sweet taste of D-asparagine and the bitter taste of L-asparagine; or the oestrogenic hormonal activity of ($+$)oestrone and the inactivity of ($-$)oestrone.

Stereoselectivity and stereospecificity

Consider the general problem of the synthesis of A_1 (a stereoisomer of A_2) starting from a pure reactant a. Many cases exist in which a is without stereoisomerism (they will be called a_0) or forms part of a set of stereoisomers $(a_1, a_2, \ldots, a_n)$:

$$a_0 \rightarrow A_1 \qquad \text{Stereoselective synthesis} \qquad (1)$$

$$a_0 \rightarrow A_1 \text{ (in excess)} + A_2 \qquad \text{Less stereoselective synthesis} \qquad (2)$$

$$a_0 \rightarrow A_1 \text{ (50\%)} + A_2 \text{ (50\%)} \qquad \begin{array}{l}\text{Absence of stereochemical} \\ \text{control in the reaction}\end{array} \qquad (3)$$

The equations (1) and (2) define stereoselective reactions; (3) represents the absence of stereoselectivity, for example the formation of a racemic compound from a prochiral compound.

An analogous system of equations exists when the starting material is a pure stereoisomer, a_1. This can give rise stereoselectively to the product A_1, to a mixture of two stereoisomers enriched in A_1, or to an equimolecular mixture of the two stereoisomers A_1 and A_2. If on starting with compound a_2 (stereoisomer of a_1), one obtains A_2, while on starting with a_1 one obtains A_1, the reaction may be called *stereospecific*.

In many reactions one stereoisomer A is synthesized as a result of the linking of several reactants (for example the Diels–Alder reaction between a dienophile and a diene).

It can be expected that an understanding of the reaction mechanism involved and its stereochemical consequences might, by a judicious choice of reactant stereoisomer a_1, a_2... allow the preparation at will of any one of the stereoisomers A of the set $(A_1, A_2, ..., A_n)$.

Another aspect of dynamic stereochemistry deals with the relative reactivity of two stereoisomers a_1 and a_2. In the extreme case one can observe the reaction $a_1 \rightarrow A_0$ (product without stereoisomerism) or $a_1 \rightarrow A_1$, without a_2 undergoing any reaction at all under the same conditions. The stereospecific enzymatic transformation of D-amino acids under the influence of D-amino acid oxidase was seen on page 92. The term stereospecific is applied here to the behaviour of the stereoisomeric reactants without consideration of the nature of the products formed.

To exemplify dynamic stereochemistry the addition reactions of olefins, the synthesis of olefins by elimination reactions, and some substitution processes will be examined. Molecular rearrangements, the reactivity of diastereoisomers, cycloadditions, and asymmetric induction will also be considered.

Among the various factors which may influence the steric development of reactions often encountered are steric effects, involving the size of substituents placed close to the reaction site; conformational effects: electronic effects, which are involved in the conditions necessary for achieving a good overlap between the orbitals of the reactant and the substrate, and which will bond them temporarily in order to lead preferentially to the transition state; and the rules of conservation of orbital symmetry (Woodward-Hoffmann rules, *see* p. 153). The effect of all these factors is sometimes described as stereoelectronic.

Kinetic and thermodynamic control

Dynamic stereochemistry has two aspects, according to whether one considers kinetically or thermodynamically controlled reactions. Kinetically controlled reactions are those in which there is no equilibrium between the stereoisomeric products formed. Thus, if stereoselectivity is observed it must be due to the difference in free energy of activation between the stereoisomeric transition states. Figure 5.1 shows the reaction in which a reactant a_0 gives stereoisomeric products A_1 and A_2 in unequal quantities:

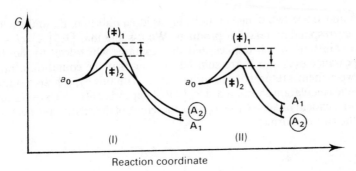

Figure 5.1

Kinetic control The stereoisomer A_2 is formed predominantly because $(G^{\ddagger})_2 < (G^{\ddagger})_1$; A_2 represents the more stable stereoisomer in case I, the less stable in case II.

Thermodynamic control
I: The stereoisomer A_1 predominates because $G_{A_1} < G_{A_2}$.
II: The stereoisomer A_2 predominates because $G_{A_2} < G_{A_1}$.

In a thermodynamically controlled reaction the experimental conditions are such that the stereoisomeric products formed interconvert and reach equilibrium. In this case, the stereoselectivity observed experimentally does not reflect the initial stereoselectivity of the reaction, it is simply a measure of the relative stability of the stereoisomers A_1 and A_2. Figure 5.1 gives the example $a_0 \rightarrow A_2 + A_1$.

For the reactions looked at from now on kinetic control will mainly be considered.

Curtin–Hammett principle

An organic compound is often conformationally heteroegeneous. If each conformation leads to a different product in a given reaction it is important to know to what extent the relative distribution of the products does or does not depend on the relative proportion of the starting conformers.

Curtin and Hammett considered the commonly encountered case in which the rate of interconversion between the conformers is much greater than the rate of the reaction (that is, the energy barrier of interconversion is much lower than the activation energy of the chemical change). Figure 5.2 illustrates the situation for the system:

$$P_A \xleftarrow{k_A} A \underset{k_{-1}}{\overset{k}{\rightleftharpoons}} E \xrightarrow{k_E} P_E.$$

A and E are two conformations in rapid equilibrium, P_A and P_E are the corresponding reaction products. We assume that $[E]/[A] = k/k_{-1}$ at any instant, which implies that the reactions which lead to the disappearance of A and E are sufficiently slow that the equilibrium ratio between them is returned to its initial value. With this hypothesis a simple kinetic calculation shows that $[P_E]/[P_A]$ depends only on $G_E^{\ddagger} - G_A^{\ddagger}$ and is independent of $\Delta G = G_E - G_A$, and, therefore, of the relative proportions of the two conformers.

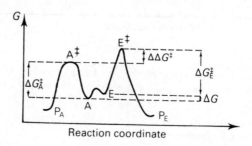

$$\Delta\Delta G^{\ddagger} = \Delta G_E^{\ddagger} = \Delta G_A^{\ddagger} + \Delta G \qquad [P_A]/[P_E] = f(\Delta\Delta G^{\ddagger})$$

Figure 5.2   The Curtin–Hammett principle

The Curtin–Hammett principle states that the proportions of the products formed are related to the difference of energy between the two transition states

$$G_E^{\ddagger} - G_A^{\ddagger} = \Delta\Delta G^{\ddagger}$$

and are independent of the position of the initial conformational equilibrium. In other words, the more reactive conformer is selected, even if it is less abundant.

The Curtin–Hammett principle often applies when the conformational changes are rotations or inversions of rings.

Some particular cases can be distinguished in Figure 5.2. It is easy to demonstrate the relationship

$$G_E^{\ddagger} - G_A^{\ddagger} = (G_E^{\ddagger} - G_E) - (G_A^{\ddagger} - G_A) + (G_E - G_A)$$

which becomes

$$\Delta\Delta G^{\ddagger} = \Delta G^{\ddagger}_{E} - \Delta G^{\ddagger}_{A} + \Delta G.$$

If by chance the two reactions have the same rate $(\Delta G^{\ddagger}_{E} = \Delta G^{\ddagger}_{A})$, the distribution of products will depend upon the initial conformational equilibrium. Also one may have equal populations of the two conformers, that is $\Delta G = 0$. The distribution of products will then be related to the difference of free energies of activation between the two reactions.

The Curtin–Hammett principle obviously does not apply when the barrier between A and E is very high. In the case of a complete hindrance of conformational exchange, the ratio of products will become identical to the relative quantities of A and E.

Addition reactions to olefins

Trans additions
Ionic additions (carried out in a polar medium with a polarizable reactant X—Y) usually lead to a *trans* addition of X and Y. A typical dissymmetric reactant such as HBr ($H^+ Br^-$) adds to 1,2-dimethylcyclohexene to give *trans*-1,2-dimethyl-1-bromocylohexane (2) (Figure 5.3):

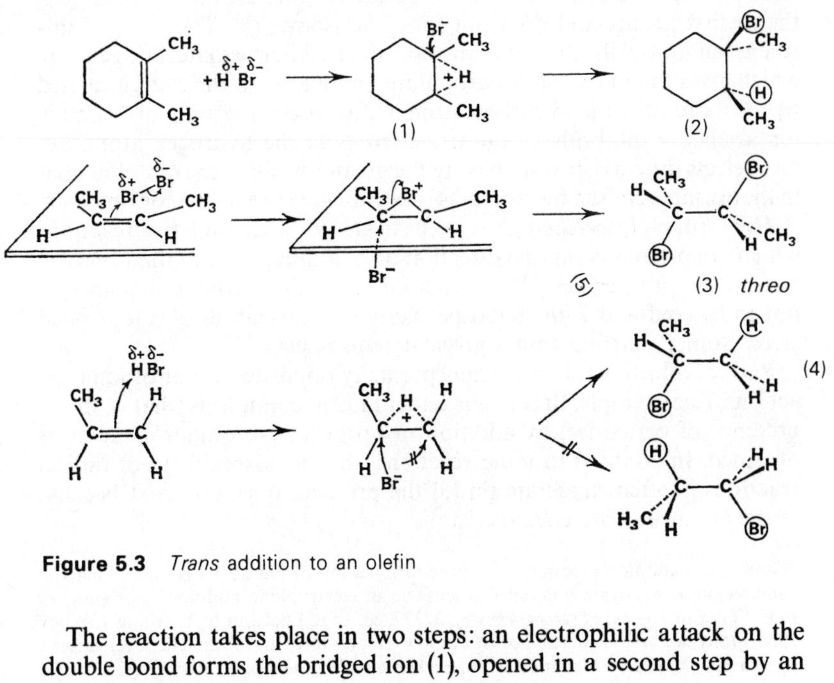

Figure 5.3 *Trans* addition to an olefin

The reaction takes place in two steps: an electrophilic attack on the double bond forms the bridged ion (1), opened in a second step by an anion which attacks the opposite side of the bridge 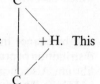 + H. This

mechanism is of a very general character[1] even in acyclic compounds. For example, bromine adding to *cis*-but-2-ene gives exclusively *threo*-2,3-dibromobutane (3) (Figure 5.3). The reaction is stereospecific. The anti conformation (3) is formed directly in the reaction, and rapidly changes to the more stable conformation.

In these two examples either the reagent or the substrate are symmetrical, which avoids any ambiguity in the structure of the product obtained. If, on the other hand, the reactant and the substrate are both dissymmetric (reagent: X—Y, olefin: $b\!\!=\!\!c$), two constitutional isomers may be formed, according to whether X becomes bonded to b or to c. Usually the two possibilities are not equivalent, one structural isomer being formed predominantly. Markownikoff's rule, applicable to hydrocarbons, states that the anionic part of the reagent becomes attached to the more substituted carbon atom. For example, $H^{\delta+}$—$Br^{\delta-}$ adds to propene giving the bromide (4) (Figure 5.3) in which Br^- is bonded to the carbon atom carrying the methyl group, and does not form the isomer (5). The basis of this rule is undoubtedly the isomerization of the intermediate bridged ion, which gives the more stable carbonium ion, that with the charge carried by the more substituted carbon atom. Otherwise expressed, the bromide ion attacking the bridged ion arrives *trans* to the hydrogen atom, but also selects the carbon atom having the greater positive charge. Note that in the example chosen the isomer (4) does not give rise to stereoisomerism.

The term *regiospecificity* has been suggested to characterize reactions which are orientated in one direction only, although more than one are possible. Regiospecificity (the formation of a single structural isomer) is not to be confused with stereospecificity (the formation of one defined stereoisomer, starting with a given stereoisomer).

Radical additions show a regiospecificity opposite to that of ionic reactions. For example, BrH reacts under radical conditions (that is, in the presence of peroxides) by addition to propene; 1-bromopropane (5) is obtained. In contrast to ionic reactions, the stereospecificity of radical reactions is often moderate (in (5) the problem does not exist because of the absence of stereoisomerism).

[1] Ionic *trans* addition is common, but the formulation of Figure 5.3 is undoubtedly as simplification; for a more detailed discussion of electrophilic addition to origins, see R. C. Fahey in *Topics in Stereochemistry*, **3**, 237, ed. E. L. Eliel and N. L. Allinger, Wiley, 1968. Many cases of *cis* addition of HCl at low temperatures (to 1,2-dimethylcyclohexene, for example) require explanation (K. B. Beckert and C. A. Grob., *Synthesis*, 1973, p. 789).

Cis additions (Figure 5.4)

Epoxidation (*cis*-butene→*cis* (6)), *cis*-hydroxylation by KMnO₄, and hydroboration will be considered. Figure 5.4 shows the synthesis of *trans*-2-methylcyclohexanol (7), starting from methylcyclohexene. The addition of boron hydride presumably takes place by a cyclic mechanism, the boron atom being the electrophile of the reagent. The oxidation of the borane intermediate with hydrogen peroxide gives the alcohol (7), without any trace of its *cis* epimer. The reaction is thus completely stereospecific (*cis* addition), and it is also completely regiospecific, the elements of water being apparently added in an 'anti-Markownikoff' orientation. This orientation is not at all anomalous, since the reagent which attacks

the double bond is not water but the hydride $H \overset{\delta-}{\underset{}{—}} \overset{\delta+}{B} {\overset{<}{}}$, and the regio-specificity is, therefore, in accord with Markownikoff's rule. The hydroboration reaction (due to H. C. Brown) is valuable because it allows the synthesis of alcohols inaccessible by the classical hydration procedures.

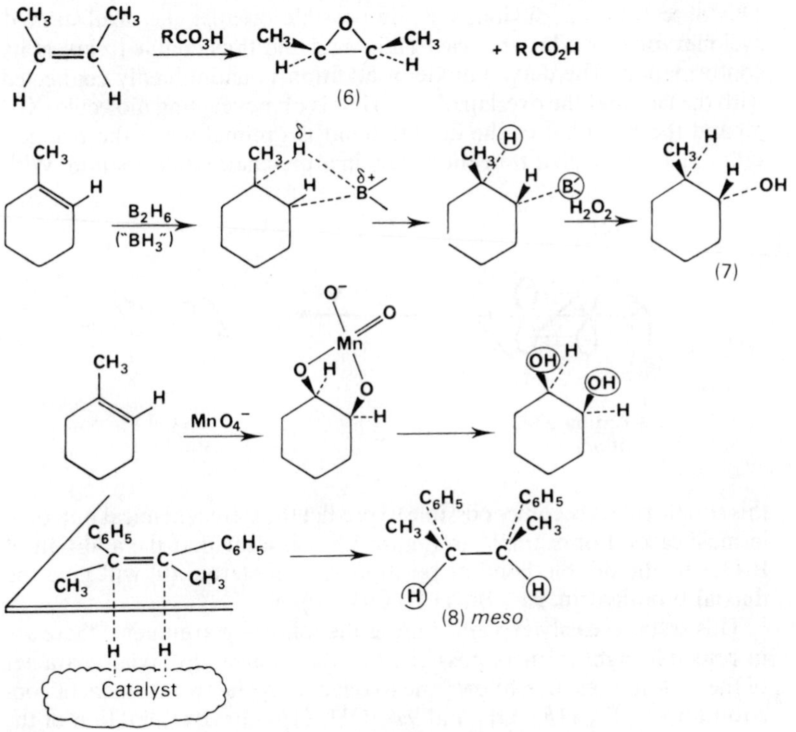

Figure 5.4   *Cis* addition to an olefin

The final example of *cis* addition is the catalytic reduction of double bonds. The catalyst (palladium, platinum or nickel) has a double role. It activates the molecule of hydrogen, which is adsorbed on its surface in the form of dissociated atoms; the metal also coordinates to the double bond of the olefin which is held on its surface. A stepwise transfer of hydrogen atoms on the same face of the double bond effects the hydrogenation (Polanyi mechanism). In Figure 5.4 a very simplified version of this mechanism is shown using 2,3-diphenylbut-2-ene, from which the *meso* diastereoisomer (8) is formed.

Cyclic systems and conformational effects

So far addition reactions have been considered only in relatively simple cyclic or acyclic systems. It is interesting now to examine rigid cyclic systems and those containing asymmetric centres. It is possible to observe the conformational aspect of the reaction of a cyclohexene ring forming part of a complex molecule, such as that of a steroid. It was Barton who first called attention to the fact that *trans* addition is essentially diaxial. This observation is obviously made possible because the disubstituted cyclohexane formed in the reaction is rigid and thus unable to invert its conformation. The diaxial mode of addition is undoubtedly connected with the fact that the overlap of the orbitals of the reacting molecule (X—Y) and the π orbital of the double bond is optimal when the reaction takes place through a transition state in a pre-chair conformation. With

Antiparallel attack

Diaxial addition

this simple rule it becomes possible to predict the stereochemical outcome in most cases. For example, in Figure 5.5 it is clear that the addition of BrOH to the double bond at position 9,11 of steroid (9) will give the diaxial bromhydrin, 9α—Br, 11β—OH (10).

This result is easily explained using the following argument; there are in principle eight isomers possible, but the *trans* and diaxial character of the addition reaction allows one to select only the two isomers of constitution 9α—Br, 11β—OH and 9α—OH, 11β—Br. Examination of the molecule shows considerable hindrance on the β face because of the presence of two angular methyl groups. The formation of the bromonium

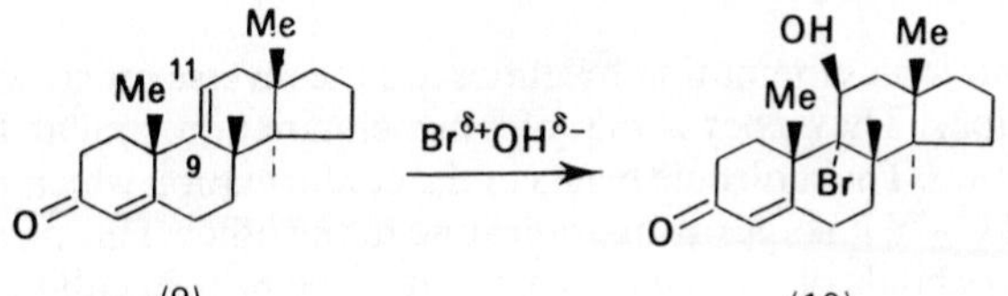

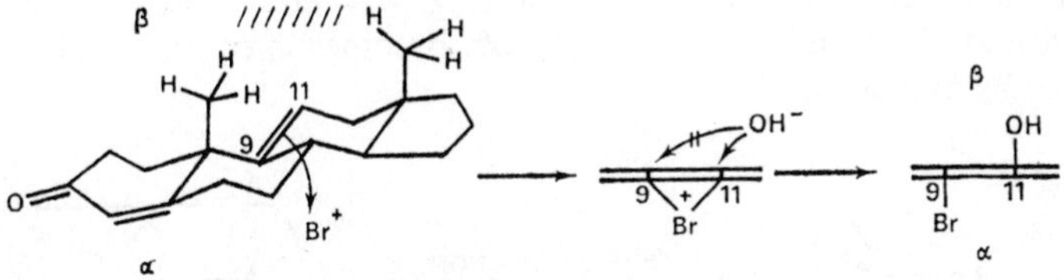

Figure 5.5

ion, which is the first step of the reaction, will therefore occur at the less hindered α face. OH^- must then necessarily approach on the β side, and will attack carbon atom 11 because this is the only way to form a diaxial bromhydrin. The regiospecificity and the stereoselectivity are thus the consequence of steric factors (preferred approach for formation of the bromonium ion) and conformational factors (diaxial addition). Note that the result is a regiospecificity opposed to that predicted by the simple application of Markownikoff's rule.

In certain cases *trans* diequatorial addition reactions have been observed; they have been explained by Valls and Toromanoff[1] as the result of a *trans* addition process perpendicular to the plane of the double bond, called 'parallel', with the development of a transition state leading to a twist ring:

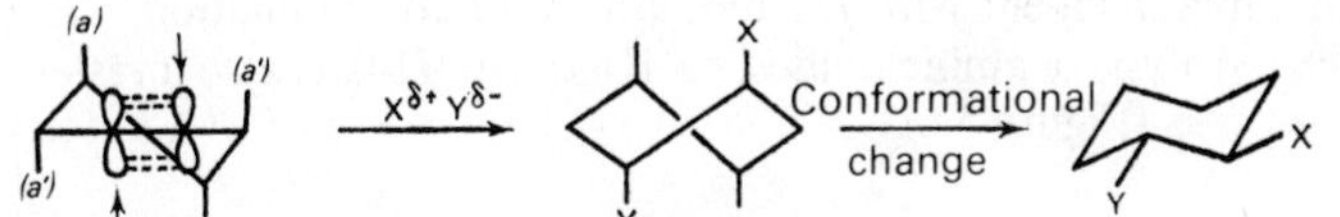

Parallel attack only takes place as a result of very special steric requirements; it is in principle much less probable than antiparallel attack because the transition state leading to the twist conformation which results is energetically unfavourable relative to its pre-chair analogue.

[1] J. Valls and E. Toromanoff, *Bull. Soc. Chim.*, 1961, 758.

Elimination reactions

Bimolecular ionic elimination reactions are the reverse processes to addition reactions. They obey a rule of *trans* elimination similar to that of *trans* addition. The molecule reacts in the conformation which places the C—X and C—Y linkages antiparallel one to the other. This arrangement makes the orbitals of the bonds which are to be broken collinear, before the final π linkage.

When the molecule cannot realize the desired *anti* conformation, for example because of the effect of a ring, *syn* elimination, in which C—X and C—Y linkages are coplanar, becomes effective as has been shown by Sicher[1]:

Figure 5.6 shows examples of *trans* and *syn* elimination.

Molecular eliminations which take place without the help of a reagent, simply by heating, are usually the result of a cyclic mechanism which favours *cis* elimination. Pyrolysis of an ester eliminates a molecule of a carboxylic acid. On pyrolysis the *erythro* diastereoisomer (11) gives *trans*-stilbene (12), after the exclusive elimination of the hydrogen atom. This double specificity is easily explained. The two possible cyclic transition states are those in which the acetoxy group eclipses H or D. In one case the two phenyl groups are *trans* one to the other (11), in the other the two phenyl groups eclipse each other, which destabilizes the corresponding transition state. The same reaction, starting from the *threo*-diastereoisomer (H and D interchanged in (11)), forms *trans*-stilbene without deuterium, in agreement with the mechanism of *cis* elimination.

The pyrolysis of amine oxides, such as (13)→(14), is also a *cis*-elimination process (Figure 5.6).

Substitution reactions

Bimolecular nucleophilic substitution (S_N2) was one of the first reactions of organic chemistry to be studied stereochemically. In 1896, Walden showed that either inversion of configuration (Walden inversion) or retention could be observed in substitution reactions, the outcome depending on the particular case. Walden's experiments are summarized in Figure

[1] J. Sicher, *Angew. Chem. Intern.*, 1972, **11**, 200.

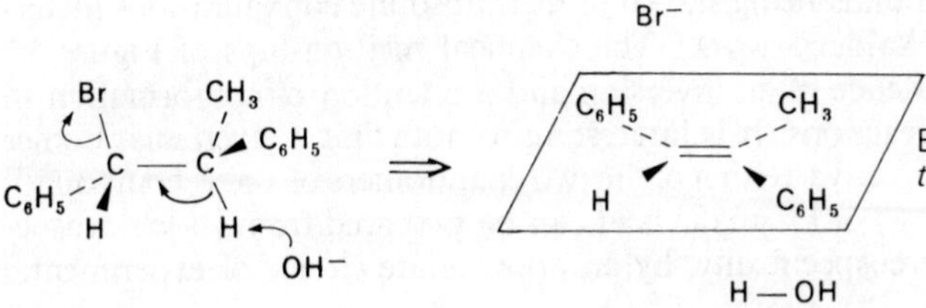

Bimolecular
trans elimination

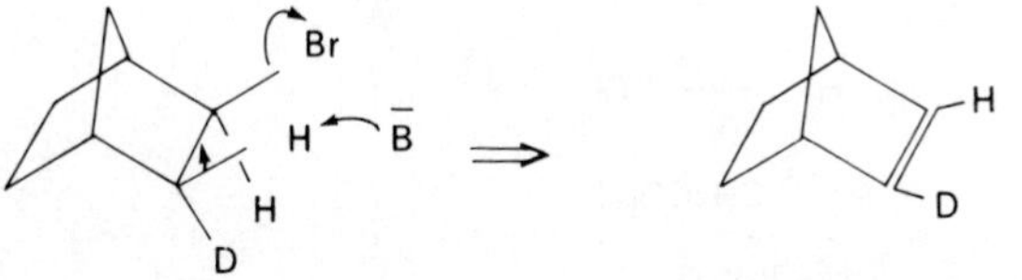

Bimolecular
syn elimination

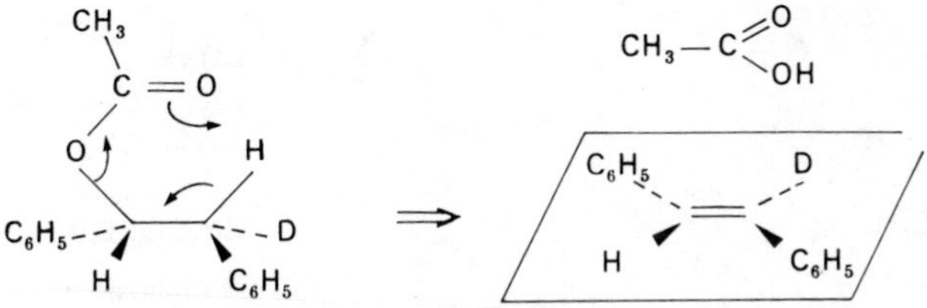

Molecular
syn elimination

(11) (12)

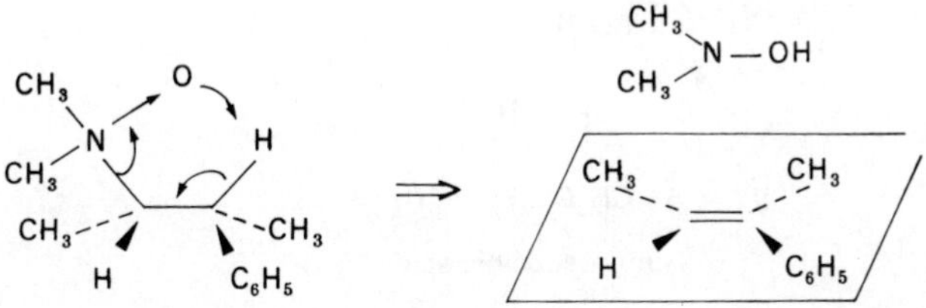

Molecular
syn elimination

(13) (14)

Figure 5.6 Elimination reactions

5.7, the compounds being shown in their absolute configurations (determined since Walden's work). The chemical relationships of Figure 5.7 prove the existence of an inversion and a retention of conformation in this series of reactions. It is interesting to note that a given enantiomer can give rise directly to either of the two enantiomers of a new compound; for example, (+) or (−)malic acid can be prepared from (+)chlorosuccinic acid, stereospecifically, by an appropriate choice of experimental conditions.

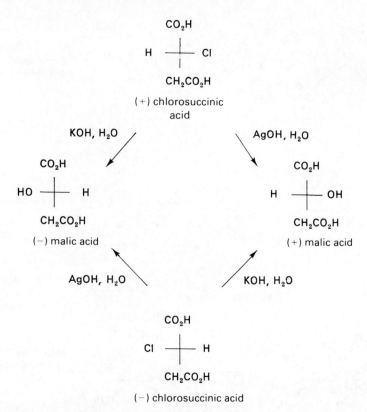

Figure 5.7   The Walden inversion

It is now known that pure S_N2 substitution always involves inversion of configuration, the transition state containing a pentacoordinate carbon atom (Figure 5.8); the leaving group X and the entering group X′ (the nucleophile) are bonded to an asymmetric carbon atom by partial bonds. The Walden inversion has long been attributed to steric factors, the reagent nucleophile X' not being hindered by the leaving group X. The symmetry of the orbitals involved may also need to be considered[1].

The reaction would then be initiated by an overlap of the lowest unoccupied molecular orbital of the substrate (an antibonding orbital as regards the C—X linkage) with the highest occupied orbital of the reacting nucleophile. The best way in which the reagent can effect an overlap allowed by the symmetry rule is by approaching *anti* to the C—X bond (Figure 5.8).

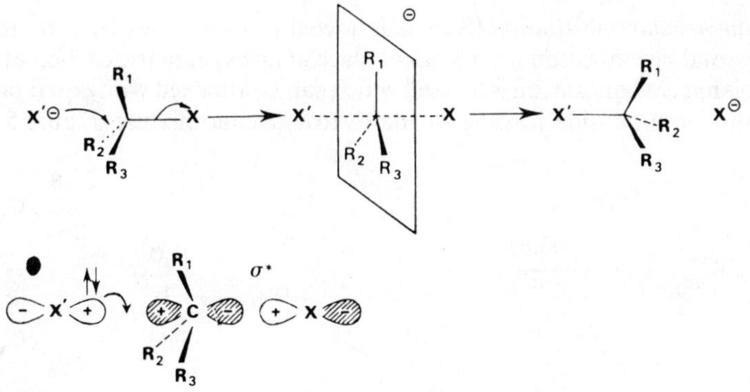

Figure 5.8 Stereochemistry of S_N2 substitution

It was mentioned on page 135 that ionic addition to an olefin involves two steps: the first is an electrophilic attack, the second a nucleophilic attack, on the triangular intermediate ion. It is easy to appreciate that

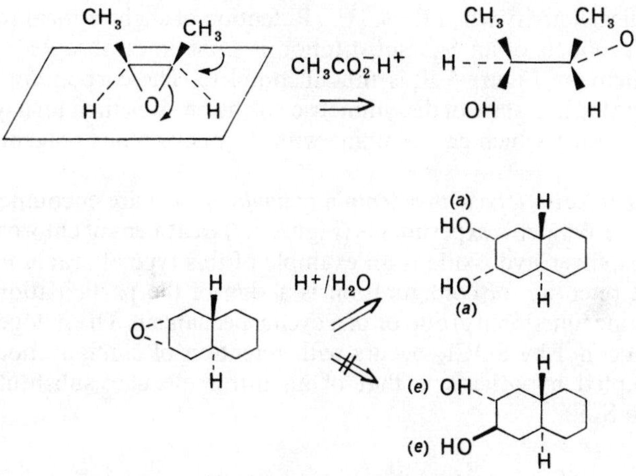

Figure 5.9 Ring-opening reactions of epoxides

[1] L. Salem, *Chemistry in Britain*, 1971, 449.

this last step is in fact a nucleophilic substitution, which involves an inversion of configuration at the carbon atom attacked. Epoxides, triangular heterocycles which can be isolated, are also ring-opened by the action of nucleophiles with inversion of configuration (Figure 5.9). The epoxycyclohexanes undergo *trans* diaxial opening, that is to say they follow the same rule as electrophilic additions to cyclohexenes. For example, hydrolysis of the epoxide (15) gives the *trans* diaxial diol (16) and not its diequatorial isomer (17).

Unimolecular substitution ($S_N 1$) is followed in most cases by a more or less total racemization when it takes place at an asymmetric carbon atom. A planar carbonium ion is formed which can be attacked with equal probability on either side, leading inevitably to a racemic mixture (Figure 5.10).

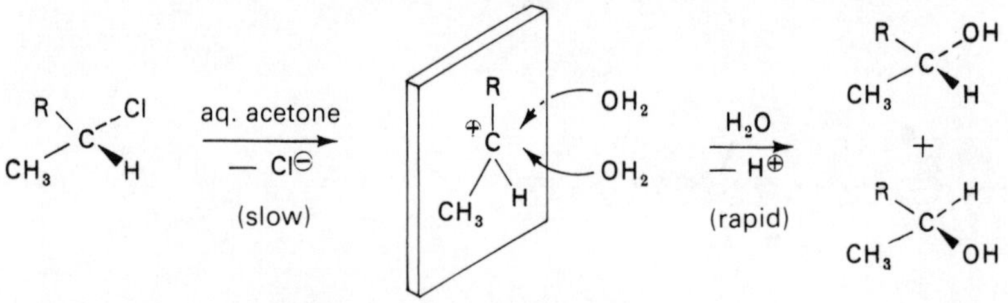

Figure 5.10   Loss of configurational integrity during an $S_N 1$ reaction

The reaction summarized by Figure 5.10 is called a solvolysis, because water plays the role of both solvent and nucleophilic reagent. Careful examination of the carbinol formed in the hydrolysis allows the detection of slight optical activity when $R = C_6 H_5$. Retention of slight optical rotation in the products of an $S_N 1$ substitution is sometimes observed; the reaction scheme of Figure 5.10 is thus incomplete. The carbonium ion can be generated in a state of dissymmetric solvation; it is then an asymmetric labile entity which can combine with the nucleophilic reagent.

Nucleophilic substitutions with retention of configuration are encountered fairly often. In Walden's experiments (Figure 5.7) treatment of chlorosuccinic acid by silver hydroxide is an example of this type of reaction. In general, the retention of configuration is a sign of the participation of a neighbouring functional group or of a cyclic mechanism. The halogenation of an alcohol by $SOCl_2$ occurs with retention of configuration; a widely accepted hypothesis is that of an intramolecular substitution (mechanism $S_N i$):

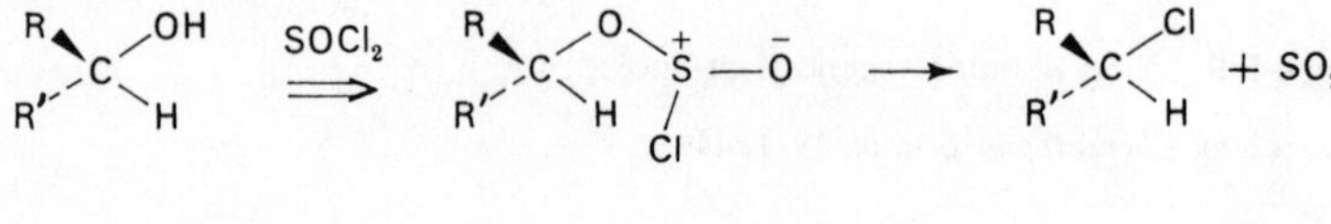

The chlorine atom cannot approach *anti* to the C—O bond because it is bonded to the sulphur atom. It is now known that this reaction is not concerted, but takes place by rearrangement of an intermediate ion pair.

Stereochemistry of molecular rearrangements

Many reactions in organic chemistry are simple molecular rearrangements of a group of atoms X moving from one point to another in the molecule. Along with these reactions we will also consider the larger group in which migration occurs but the final product is not an isomer of the starting material.

Some examples of rearrangements will be considered from the point of view of their stereochemistry. It is essentially by comparing the initial and final stereochemistry at the sites X, *a* and *b* that most information can be drawn about the mechanism and the steric course of the reaction studied.

1,2 Rearrangements in which a radical X moves from one carbon atom (centre *a*) to a heteroatom (centre *b*) are well known. In general, if the radical X is asymmetric the rearrangement occurs with retention of configuration at X. The migration of a nucleophilic radical X to an electrophilic nitrogen atom constitutes one step in the Beckmann, Curtius, Hofmann and Schmidt reactions. Figure 5.11 illustrates the mechanism of the Hofmann degradation, which converts an amide to an amine with the loss of a carbonyl group. Treatment with a hypobromite converts the amide to the N-bromoamide, which then decomposes to an acylnitrene. In this neutral intermediate the nitrogen atom is electron-deficient, and this stimulates the migration of the neighbouring group. It has been shown that the radical which migrates completely conserves its stereochemical integrity. The same is true in the Beckmann rearrangement, where one also observes great stereospecificity for *anti* migration (it is the group *trans* with respect to the hydroxyl grouping which migrates). Migration to an electron-deficient oxygen atom is also possible; it is the important step of the Baeyer–Villiger reaction, which transforms a ketone to an ester by formal insertion of an oxygen atom α to the carbonyl group. This reaction takes place as a result of the action of a peracid (Figure 5.11). The migratory aptitude of the methyl group is low, but this will not be discussed here.

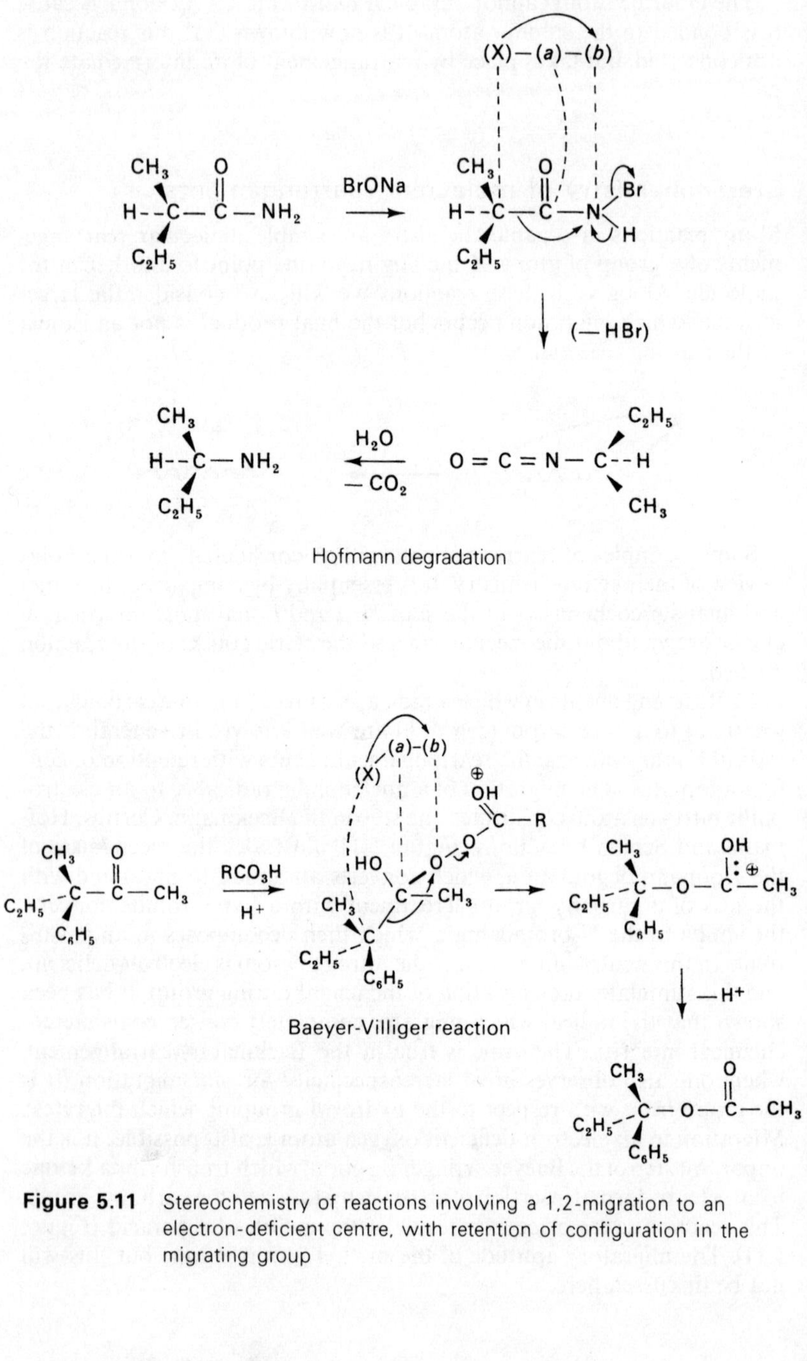

Hofmann degradation

Baeyer-Villiger reaction

Figure 5.11   Stereochemistry of reactions involving a 1,2-migration to an electron-deficient centre, with retention of configuration in the migrating group

All these examples concern the migration of a chiral group. Stereospecificity can also be associated with the acceptor centre (b) if this is a carbon atom; for example, the spinal rearrangement of steroids by a series of 1,2 migrations of hydrogen atoms and angular methyl groups which all remain on the side of the plane of the molecule where they originally were.

Asymmetric induction and asymmetric synthesis

When an asymmetric carbon atom is formed in a chemical reaction, two different situations (a) and (b) can arise if an asymmetric centre is already present in the system. Consider the reduction of a prochiral ketone by a hydride. If an asymmetric centre of configuration S is present in the ketone, the reaction may be summarized thus:

The diastereoisomers (S) ⤳ (R) and (S) ⤳ (S) can be formed in different proportions. Therefore, there is asymmetric induction by the asymmetric centre (S) in the reduction. Now consider a prochiral ketone not containing an asymmetric carbon atom, but assume the use of a hydride which is itself optically active [for example LiAlH(O—$\overset{*}{C}$H—C$_6$H$_5$)$_3$].
$$\underset{CH_3}{|}$$

The following reaction (*b*) will take place:

b)

$$X - C\overset{O}{\underset{Y}{\diagup}} + {}^{"}H"\rightsquigarrow (S) \longrightarrow X - \underset{Y}{\overset{OH}{C}}\text{--}H + X - \underset{Y}{\overset{H}{C}}\diagdown OH + (S) \rightsquigarrow$$

$$\qquad\qquad\qquad\qquad\qquad\qquad (R) \qquad\qquad (S)$$

H $\rightsquigarrow$ (S) symbolizes an optically active hydride in which the asymmetric carbon atom possesses an absolute configuration S. If the two alcohols (R) and (S) are formed in different proportions, asymmetric induction has again taken place. The inducing centre (S) in this case is in the reagent and is removed in the final treatment with water and acid, because it is not bonded to the synthesized alcohol.

These examples (*a*) and (*b*) represent the two methods for preparing an asymmetric centre in a given absolute configuration. In example (*a*) the asymmetric induction exercised by (S) determines the preferential attack on one of the diastereotopic faces of the prochiral ketone. Izumi[1] has suggested the name diastereoselective for this type of synthesis. One diastereoisomeric alcohol is formed predominantly. If the chiral part (S) is detachable, the alcohol X—CHOH—Y is obtained with predominance of one enantiomer, and the total process is called an asymmetric synthesis.

Scheme (*b*) is related to the preferred reaction of one of the enantiotopic faces of a prochiral substrate. The reaction is called enantioselective, and leads to direct asymmetric synthesis of X—CHOH—Y; the inducing fragment (S), being external to the substrate, does not end up bonded to the alcohol obtained.

Models of asymmetric induction

If the chiral part of the molecule responsible for the asymmetric induction is bonded to the prochiral carbon atom at which the reaction takes place, it is sometimes possible to predict the absolute configuration of the asymmetric carbon atom formed. The prediction is relatively easy when the chiral molecule is cyclic. The simple examination of molecular models often allows us to detect which diastereotopic side is the less encumbered, and therefore the more reactive. This argument is used on pages 138–9 for the steroid series; reactions on the lower or α face are generally preferred. When the prochiral centre is not situated in a ring, models of asymmetric induction become necessary. They are used to compare the relative stabilities of the two diastereoisomeric transition states. Typically, a geometry is postulated for the transition state and a hypothetical geometry

[1] Y. Izumi, *Angew. Chem. Intern.*, 1971, **10**, 871.

or structure around the reaction centre is assumed. This hypothesis remains debatable; it can only be valid when the reactions are rapid and exothermal[1]. The suggested models often have good predictive value but remain essentially empirical in character and should be considered as working rules. Two such classic rules, which will be discussed here, are those of Cram and Prelog.

Cram's rule allows the prediction of the stereochemistry for the formation of one asymmetric centre next to another asymmetric centre (1,2 induction). Consider a carbon atom C (L,M,S) where L, M, S signify groups of little polarity, unable to coordinate with an organometallic atom. L designates the largest group and S that which is the least encumbering. L is placed furthest from the carbonyl group which complexes with the organometallic compound m—R'. The radical R' attacks the carbonyl at the less hindered side, that is avoiding M:

Predominant diastereoisomer

Rules similar to that of Cram, but allowing more quantitative interpretation of the observed stereoselectivity, have been proposed. These will not be considered here[2].

Prelog's rule is useful for predicting the direction of 1,4 asymmetric induction in asymmetric synthesis.

The absolute configuration of the hydroxyacid formed (atrolactic acid) is evidently related to that of the optically active alcohol (HOC (L,M,S) used as starting material. Prelog has rationalized the experimental results by means of a scheme for asymmetric induction which postulates a conformation for the reaction in which the two carbonyl groups are antiparallel and the ester function planar. Furthermore, the large group L

---

[1] As a consequence of Hammond's principle, which will not be considered in detail here. The statement of this principle was made by Hammond in *J. Amer. Chem. Soc.*, 1955, **77**, 334.

[2] For a modification of Cram's rule *see* M. Cherest, H. Felkin and N. Prudent, *Tetr. Letters*, 1968, 2199.

$$C_6H_5-\underset{\underset{O}{\|}}{C}-CO_2H \;+\; OH-\underset{\underset{M}{\uparrow}}{C}\!\!\overset{L}{\underset{S}{\diagdown}} \;\longrightarrow\; C_6H_5-\underset{\underset{O}{\|}}{C}-\underset{\underset{O}{\|}}{C}-O-\underset{\underset{M}{\uparrow}}{C}\!\!\overset{L}{\underset{S}{\diagdown}}$$

$$\Big\uparrow \qquad\qquad\qquad \Big\downarrow CH_3MgX$$

$$C_6H_5-\underset{\underset{CO_2H}{}}{\overset{OH\quad CH_3}{C}} \;\xleftarrow{\quad OH^-\quad}\; C_6H_5-\underset{\underset{O}{\|}}{\overset{OH\quad CH_3}{C}}-C-O-\underset{\underset{M}{\uparrow}}{C}\!\!\overset{L}{\underset{S}{\diagdown}}$$

is assumed to lie in the plane of the two carbonyl groups. Under these conditions the prochiral carbonyl group is attacked on its less hindered face, that containing the group S. Prelog's rule has been widely used for determining the absolute configuration of a chiral alcohol by examining the configuration of the atrolactic acid formed in the asymmetric synthesis.

$$C_6H_5\overset{O}{\underset{O}{\overset{\|}{\diagdown}}}\;\overset{S}{\underset{L}{\overset{\|}{\diagup}}}M \;\longrightarrow\; \underset{CH_3}{\overset{C_6H_5}{\diagup}}\underset{OH}{\overset{CO_2H}{\diagdown}}$$

$$CH_3MgX$$

R(+)-atrolactic acid

Cycloadditions

Cycloadditions are reactions in which several molecules combine to form a ring. The best-known cycloaddition is the Diels–Alder reaction, or diene synthesis, which involves heating a diene and a monoolefin (called a dienophile). In Figure 5.12 the condensation of budadiene and dimethyl maleate is shown. The reaction is completely stereospecific and gives a *cis*-dimethyl tetrahydrophthalate. The diene can be regarded as a reagent effecting a *cis* addition to the double bond of the dienophile. The *cis* or *trans* stereochemistry of this reactant is retained in the final product. Additional information on the steric course of the Diels–Alder reaction is provided by the condensation between *trans,trans*-1,4-dimethylbuta-diene and dimethylmaleate (Figure 5.12). The exclusive product of the cycloaddition has the stereochemistry indicated (18) so that the dieno-phile has attacked the butadiene *endo*, the ester groups being placed on the side of the original diene system, and not opposite, as one might have predicted. This *endo* attack is very general and presumably corresponds to secondary interactions, favourable for the reaction, between the two conjugated double bonds and the functional groups of the dienophile. The two methyl groups present in the butadiene also conserve a relative *cis* stereochemistry in the product of the cycloaddition reaction. The product thus retains the 'imprint' of the *trans, s-cis, trans* stereochemistry in the initial butadiene conformation, and testifies to the *cis* addition of the dienophile and the diene system, assumed to be more or less planar.

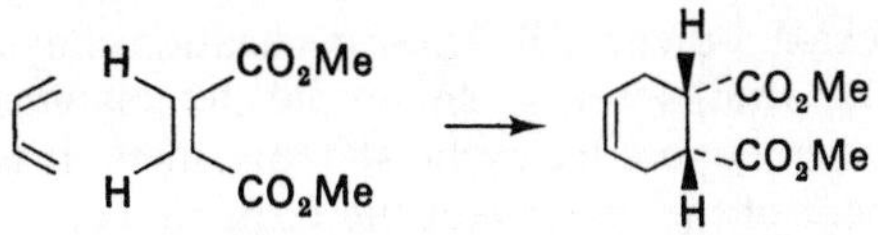

Diene   Dienophile

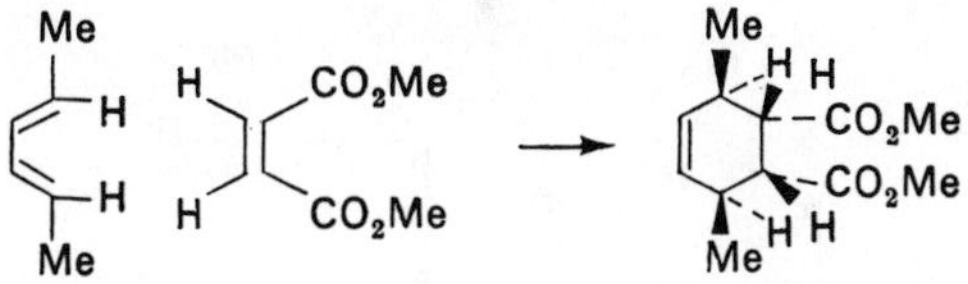

(18)

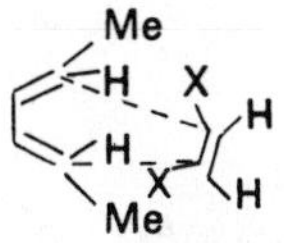

(19)

Figure 5.12

All these stereochemical results show that the transition state of the Diels–Alder reaction can be represented to a good approximation by (19).

It is interesting that the adduct (18) contains four centres of asymmetry. A nonstereospecific cycloaddition reaction would give eight different racemates. The stereochemistry of the one racemate which is actually obtained is easily predicted if one remembers that the reaction has a character of *cis* overlap (with respect to both the diene and the dienophile) and an *endo* specificity.

Cyclization and electrocyclic reactions

Many intramolecular cyclizations of conjugated polyenes, taking place thermally or by a photochemical route, are completely stereospecific. These reactions (or their inverse) are called electrocyclic reactions; for example, the photochemical cyclization of butadienes and of 1,3,5-hexatrienes (Figure 5.13).

The *trans,trans*-butadiene (20) is cyclized to *cis*-3,4-dimethylcyclobutene (21), whereas the *trans,cis,trans*-hexatriene (22) gives *trans*-5,6-dimethyl cyclohexa-1,3-diene (23). These cyclizations have a well-defined steric course which results in a given acyclic stereoisomer yielding only one of the theoretically possible cyclic stereoisomers. It is usual to consider the movement of the groups on the terminal carbon atoms which become bonded to each other in the course of the cyclization. Without knowing in detail the geometry of the transition state it is easy to see the direction of the movement of the methyl groups of (20) and (22). A

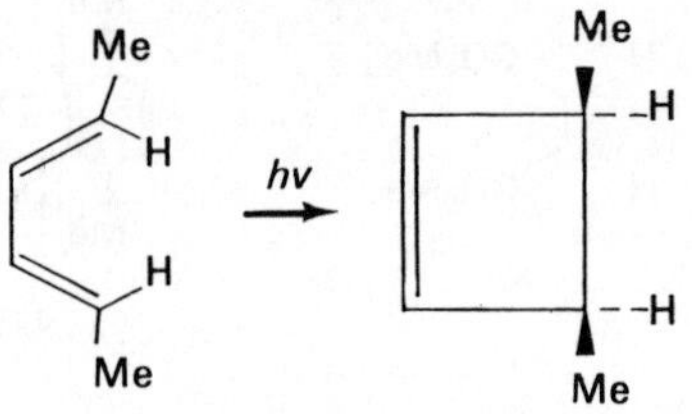

(20) (21)

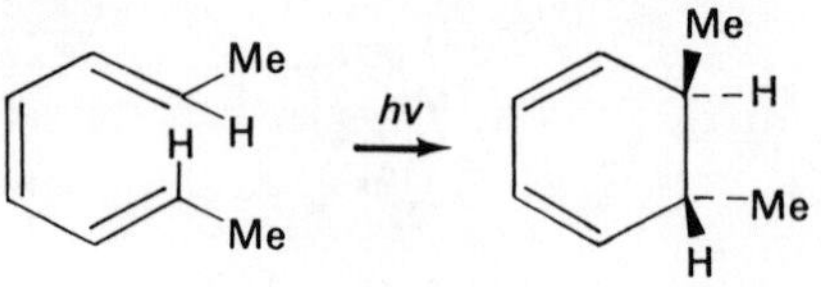

(22) (23)

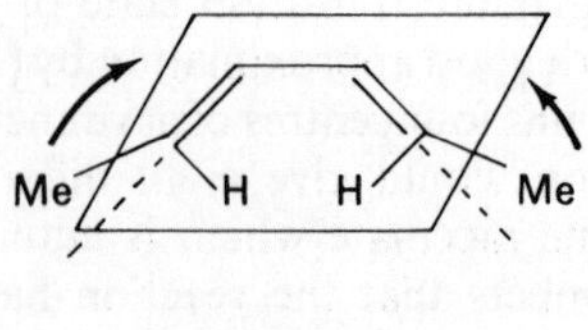

Disrotatory cyclization

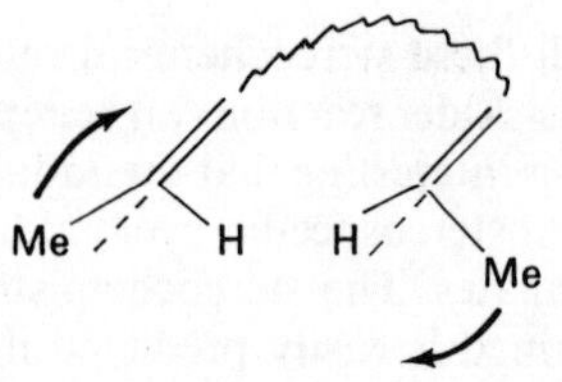

Conrotatory cyclization

Figure 5.13

disrotatory cyclization (rotations in opposite directions) necessarily occurs in the transformation of (20)→(21). In contrast, a conrotatory cyclization (rotations in the same direction) is necessary to effect the transformation (22)→(23).

These few facts were selected from an enormous mass of experimental results to show that the majority of concerted reactions (reactions which take place in a single step, and in which the rupture and the formation of bonds are not independent) obey strict stereochemical rules. It was Woodward and Hoffmann, in 1965, who clarified and rationalized a situation which until then was very complex. They proposed simple rules[1] allowing the prediction of stereochemical consequences. These rules have been established by considering the symmetry of the molecular orbitals involved in these reactions.

Principle of the Woodward–Hoffmann rules

It is not possible in the framework of this book to go deeply into this problem; it will merely be touched upon. Only cyclization reactions will be considered. Woodward and Hoffmann suggested taking into consideration the symmetry of the highest occupied molecular orbital in thermal cyclizations. In photochemical cyclizations it is the symmetry of the lowest unoccupied molecular orbital which is decisive for the prediction of the steric course of the cyclization. The importance of knowing the symmetry of a molecular orbital must be considered.

The terminal atoms are destined to become bonded in forming the ring, and the final σ bond is formed by the overlap of two orbitals of the same sign. The conrotatory or disrotatory character of the reaction depends upon the relative position of the orbitals of the same sign:

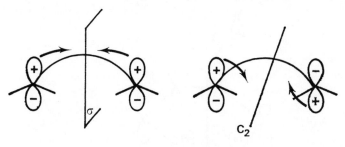

Disrotatory
(movement symmetrical
about a plane)

Conrotatory
(movement symmetrical
about a C_2 axis)

[1] R. B. Woodward and R. Hoffmann, *Conservation of Orbital Symmetry*, Academic Press, 1970.

With these very simple ideas it should be easy to predict whether a given reaction should be disrotatory or conrotatory. Consider the cyclization of a butadiene to a cyclobutene (Figure 5.14):

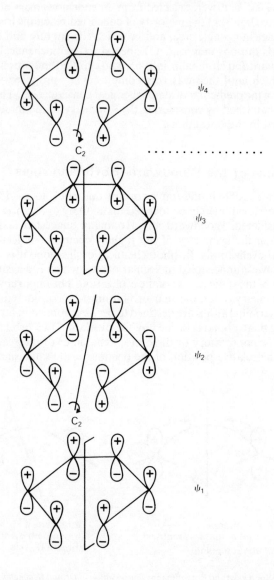

Figure 5.14

The four molecular orbitals of butadiene, Ψ_1, Ψ_2, Ψ_3, Ψ_4, were shown in Chapter 1 (p. 31). The highest occupied molecular orbital is the orbital Ψ_2, containing an axis of symmetry. Clearly cyclization by means of Ψ_2, in a thermal reaction, must have a conrotatory character. The thermal reaction is in fact difficult, although the photochemical reaction is easy. The lowest unoccupied molecular orbital, Ψ_3, should be considered in this case. This contains a plane of symmetry and will favour disrotatory cyclization.

If the conjugation is extended by adding another double bond the situation will change. The lower molecular orbitals of a hexatriene are shown in Figure 5.14. The photochemical cyclization is directed by the symmetry of Ψ_4, the lowest unoccupied molecular orbital. Ψ_4 contains an axis of symmetry, and the cyclization is therefore conrotatory, in agreement with experiment.

Sigmatropic rearrangements

A rearrangement which has the effect of changing the relative position of adjacent double and single bonds is called a sigmatropic rearrangement. For example, the Cope rearrangement is a concerted process which takes place on heating:

(24)   (25)

This reaction, allowed by a thermal pathway, shows marked stereospecificity. The *erythro* isomer (24) gives almost exclusively the *cis,trans*-diene (25) on pyrolysis.

The concerted reaction goes through a cyclic transition state. Assuming that the transition state has a chair, rather than a boat, conformation (molecular orbital calculations agree with this interpretation), one correctly predicts the stereochemistry of the final product:

Rearrangements in which a migration of a hydrogen atom takes place are also sigmatropic rearrangements. On heating, calciferol (vitamin D_2) is converted to precalciferol, the hydrogen atom at 9 being transferred to carbon atom 19:

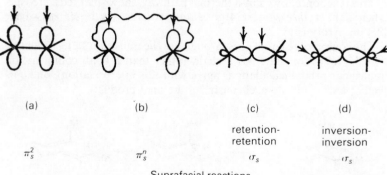

Calciferol   Precalciferol

According to the terminology of Woodward and Hoffmann this is a (1,7) sigmatropic transposition.

Pericyclic reactions

Woodward and Hoffmann have suggested grouping under the general name 'pericyclic reactions' those concerted reactions which have the common characteristic of a cyclic transition state: i.e. cycloadditions, electrocyclic reactions, sigmatropic reactions; the total number of electrons involved giving an indication of the stereochemistry of these reactions through some very general rules, summarized below.

Antarafacial or suprafacial character

The terms antarafacial (a) and suprafacial (s) were introduced by Woodward and Hoffmann to denote the stereochemistry of a reaction with reference to one face of a system of π electrons or orbitals. Antarafacial implies that both faces are involved in the reaction, suprafacial implies a process which involves only one face of the system.

(a)	(b)	(c)	(d)
		retention-retention	inversion-inversion
π_s^2	π_s^n	σ_s	σ_s

Suprafacial reactions

(e) (f) (g)

inversion-retention

$\bar{\pi}_a^2$ $\bar{\pi}_a^n$ σ_a

Antarafacial reactions

The definition of the terms supra/antara has been generalized to reactions involving σ bonds themselves. The various possibilities for the cleavage of a σ bond with the stereochemical consequences of inversion or retention of configuration are indicated schematically.

The designation supra/antara contains stereochemically important information; it states which lobes of the orbitals are used in forming the bonds. The Diels–Alder reaction is classed as a suprafacial reaction with respect to the diene (there is certainly *cis* addition at the extremities of the diene, as we have seen on p. 150; but *see also* case (b), above). The Diels–Alder reaction is similarly suprafacial for the dienophile, (a). The stereochemical facts are summarized by saying that the diene synthesis is of the type $\pi_s^4 + \pi_s^2$ (π^4 and π^2, recalling that π systems with 4 and 2 electrons respectively are involved. The photochemical cyclization of butadiene to cyclobutene (p. 152), which has a disrotatory character, is also of the suprafacial type, (b). A concerted *trans* addition to a double bond is of the type π_a^2, (e). Note that the great majority of ionic additions take place in two steps (p. 140), the electrophile carrying out a *cis* addition in the first step (addition π_s^2, (a)). The *trans* addition which results is in fact an attack on the σ bond carried out with an inversion of configuration on the carbon atom substituted (reaction π_a^2, (g)). The classical S_N2 reaction is similarly an antarafacial process with respect to the carbon atom which is the point of substitution.

The migration of a radical (hydrogen for example) during a rearrangement can be suprafacial or antarafacial. For example, the Woodward–Hoffmann rules predict that the thermal rearrangement (1,5) shown below is suprafacial (with respect to the ring, (b)):

When a radical migrates it is always important to consider its possible change of configuration. It is known that in the Wagner–Meerwein reactions the group which migrates retains its configuration; this is general for 1,2 migrations:

The reaction is suprafacial with respect to the carbon atom which migrates, (c).

These few examples exemplify the diversity of reactions in which the stereochemistry of a concerted process (thermal or photochemical) is imposed. Rules have been given for rapidly predicting their steric course.

Selection rules for pericyclic reactions

In considering the conservation of orbital symmetry, Woodward and Hoffmann have demonstrated a very general rule for pericyclic reactions:

A pericyclic reaction is thermally allowed if the total number of elements $(4q+2)_s$, used in a suprafacial manner, and $(4r)_a$, used in an antarafacial manner, is odd.

This number is even in photochemical reactions. $4q+2$ and $4r$ are the number of electrons which are present in the elements considered as taking part in the reaction.

This rule can immediately be applied to the thermal cyclization of butadiene:

Butadiene contains 4π electrons, that is a 4π system. Butadiene, therefore, contains an element $4r(r=1)$. If this element is used in an antarafacial fashion (conrotatory reaction) the reaction is thermally allowed. The cyclization of 1,3,5-hexatriene can be treated in the following way:

The triene constitutes a single element of type π^6, which contains $4q+2$ electrons ($q=1$) and no $4r$ element. The total is odd, so the thermal reaction is possible with a π_s element (suprafacial reaction, or disrotatory).

The Diels–Alder reaction is a reaction with two elements ($\pi^2 + \pi^4$):

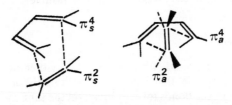

$$\pi^4 \quad \pi^2$$

In order that a thermal reaction is possible it is necessary to have an odd number of elements $(4q+2)_s$ and $(4r)_a$. The elements $(4q+2)_a$ and $(4r)_s$ are not taken into account for the application of the generalized Woodward–Hoffmann rule. In the combination $\pi_s^2 + \pi_a^4$ there is an even number of 'useful' elements and the thermal reaction cannot be carried out. If the Diels–Alder reaction is classed as $\pi_s^2 + \pi_s^4$ it becomes allowed thermally (1 element $(4q+2)_s$, 0 element $(4r)_a$). Another way of being in accordance with the rule is to consider the combination $\pi_a^2 + \pi_a^4$ (0 element $(4q+2)_s$, 1 element $(4r)_a$):

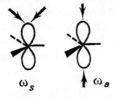

This last mode of reaction implies an unfavourable orthogonal approach of the reactants; it is the process $\pi_s^2 + \pi_s^4$ which is preferred and which is observed experimentally. This method of analysis will not be developed further; the basic principles only have been given. For further detail the reader should refer to more specialized work[1]. It must be mentioned, however, that the methods used for counting the elements have a somewhat arbitrary character (for example, π^4 could be also considered as $\pi^2 + \pi^2$). The elements composed of a single orbital (empty or containing a doublet) could also be counted, with an antarafacial or

$$\omega_s \qquad \omega_a$$

[1] R. B. Woodward and R. Hoffmann, *Conservation of Orbital Symmetry*, Academic Press, 1970.

suprafacial character. These orbitals are named ω (nonbonding doublet, carbonium ion, carbanion...).

For example, the Wagner–Meerwein rearrangement (p. 158) is of the type $\sigma^2 + \omega^0$. There is an element $4q + 2(\sigma^2)$ and an element $4r(\omega^0)$. In order that the reaction shall be thermally allowed one must only retain a single element: $(4q + 2)_s$ or $(4r)_a$. Otherwise expressed, two solutions exist in theory: $\sigma_s^2 + \omega_s^0$ and $\sigma_a^2 + \omega_a^0$. The latter case is geometrically impossible, and there only remains suprafacial migration with retention of configuration at the centre of the group which migrates.

Parity of a reaction

The steric course of most concerted reactions can be predicted very simply by counting the number of doublets involved and applying the parity rules due principally to Mathieu[1] and to Rassat[2]. The method of presentation introduced by Mathieu, which covers most known reactions, will be used here.

Even reactions are those in which an even number of doublets is involved, odd reactions use an odd number of doublets.

Rule of parity

In even reactions the bonds form or break on either side of the plane of the reacting molecule, that is to say in both half-spaces.

In odd reactions the bonds form or break on the same side of the reacting molecule, that is to say in the same half-space.

When a reaction occurs at several stereochemical sites it is possible to divide it up, site by site, in such a way that the product of the local stereochemistries shall be equal to the overall stereochemistry[3].

These rules can be stated simply by saying that even reactions have a generally antarafacial stereochemistry, whereas odd reactions are characterized by suprafacial stereochemistry (Figure 5.15).

The application of these rules to concerted reactions is extremely simple and rapid. Some examples are given below.

Consider the closure of butadiene to cyclobutene, discussed on page 152. This is an even reaction (two doublets), therefore the thermal cyclization should be conrotatory. The diene synthesis (p. 151) is an odd reaction (three doublets), it should have an overall supra character, which is obtained by a suprafacial (or *cis*) stereochemistry of addition of each partner.

[1] J. Mathieu, *Bull. Soc. Chim.*, 1973, 807; *C.R. Acad. Sci.*, 1972, **C, 274**, 81.
[2] A. Rassat, *C.R. Acad. Sci.*, 1972, **C, 274**, 730.
[3] $(\text{retention})^{(s)} \times (\text{inversion})^{(s)} = (\text{inversion})^{(a)}$ ...

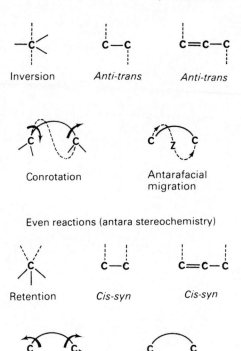

Inversion   *Anti-trans*   *Anti-trans*

Conrotation   Antarafacial migration

Even reactions (antara stereochemistry)

Retention   *Cis-syn*   *Cis-syn*

Disrotation   Suprafacial migration

Odd reactions (supra stereochemistry)

Figure 5.15

The 1,2 migration of a carbon radical to an electron-deficient atom (p. 158) is an odd reaction (one doublet), and the reaction should therefore take place with retention of configuration at the carbon centre which migrates. The elimination reactions (11)→(12) and (13)→(14) (Figure 5.6) involve the use of three doublets, and are therefore odd reactions which should have suprafacial stereochemistry, which is in accordance with the *syn* elimination observed. Finally, the nucleophilic bimolecular substitution reaction S_N2 involves two doublets at the carbon atom where the reaction occurs; the reaction is therefore even and should have an antarafacial character, and this is confirmed by the inversion of configuration which takes place.

Index

Index of names